MUSKEGON AREA DIS LIBRARY (MADL)
3 1368 00 5

BATS SING MICE GIGGLE

REVEALING THE SECRET LIVES OF ANIMALS

Karen Shanor and Jagmeet Kanwal

ICON BOOKS

Published in the UK in 2009 by
Icon Books Ltd, Omnibus Business Centre,
39–41 North Road, London N7 9DP
email: info@iconbooks.co.uk
www.iconbooks.co.uk

Sold in the UK, Europe, South Africa and Asia
by Faber & Faber Ltd, Bloomsbury House,
74–77 Great Russell Street, London WC1B 3DA
or their agents

Distributed in the UK, Europe, South Africa and Asia
by TBS Ltd, TBS Distribution Centre, Colchester Road,
Frating Green, Colchester CO7 7DW

Published in Australia in 2009
by Allen & Unwin Pty Ltd,
PO Box 8500, 83 Alexander Street,
Crows Nest, NSW 2065

ISBN 978-184831-071-1 (hardback)
ISBN 978-184831-095-7 (paperback)

Typesetting in 11pt Plantin by Marie Doherty

Printed and bound in the UK by
CPI Mackays, Chatham

Contents

Karen Shanor is a neuropsychologist, a former White House consultant and an advisory member of Discovery Channel Global Education. At Stanford University she researched how rats learn, and how cats dream. Her work at NASA's Life Sciences department included animal research on memory and information theory, and she has taught with Karl Pribram at Georgetown University since 1998. As a Peace Corps science teacher in Somalia, she was a consultant for a wildlife conservatory. A frequent lecturer at the Smithsonian Institution, Karen also hosted an NBC radio programme for five years and appears frequently on *Larry King Live*, *CBS Nightly News*, *Dateline*, *The Today Show* and *Oprah*, and is a regular contributor to CNN.

Jagmeet Kanwal, Ph.D is an associate professor in the Department of Physiology and Biophysics and the Department of Psychology at Georgetown University in Washington, DC. He is also an external professor at the Krasnow Institute for Advanced Study in Fairfax, Virginia. Dr Kanwal is an internationally recognised neuroethologist who was the first to perform magnetic resonance imaging in awake animals. He is an expert on cortical mechanisms for the perception of complex sounds. Dr Kanwal discovered a left-brain dominance for species-specific calls in bats, as is present for speech in humans, and together with co-workers is engaged in cracking the code for the neural representation of social calls. Dr Kanwal's early contributions on the comparative organisation of chemosensory systems include the discovery of taste centres in the forebrain of fish. He uses interdisciplinary approaches to understand the functional organisation of the brain from the viewpoint of behaviour. He is also an ardent birdwatcher and keen nature photographer.

Both authors live in the Washington, DC area.

Acknowledgements

First and foremost, we are grateful to Icon editor Simon Flynn and his extraordinary group, especially Nick Sidwell, Andrew Furlow and Duncan Heath, who laboured with patience night and day across multiple time zones to produce a book with cutting-edge science. Their understanding and skill in coordinating the tasks that needed to be performed at various stages of the production were outstanding. We thank our agents, Muriel Nellis and especially Jane Roberts of the Literary and Creative Artists, Inc. for their expertise and constant support in this endeavour. Jane's immense enthusiasm and literary wisdom kept us feeling optimistic about completing the book during difficult times. She was always there for us when we needed a quick second opinion or a reader's perspective. On the academic side, we are deeply indebted to Karl Pribram, who at the age of 90 continues to be a great source of inspiration. His engaging lectures and stimulating discussions at Georgetown University provided an intellectual forum for many thoughtful interactions between the authors. We are also thankful to Georgetown Professor Patrick Heelan for his guidance on the quantum physics and seismological concepts alluded to in this book. John Caprio, Thomas

Finger and Nobuo Suga also have been scientific mentors whose interesting research has contributed to some of the findings reported in this book.

Marine biologist Robert Woollacott of Cambridge, Massachusetts, Ken Ferebee of the US National Park Service, Stuart Brown, author of the book *Play*, mathematician James Shanor, theoretical physicist Sarbmeet Kanwal, geologist Gordy Shanor, and David Wood of the Sidwell Friends science programme all shared their professional expertise that helped to improve earlier drafts of the manuscript. We also thank Ian Hay Falconer and Constance Culler Falconer for their literary research, and Daniel Perry and Maxinder Kanwal for their critical input to the manuscript. Walt Ellison provided his invaluable computer expertise to keep our communication lines buzzing across the two continents.

We also want to acknowledge Srimati Kamala, Nancy Bugos, Muthiah Veerappan, Niranjan M. Shah, Madhav Singh Parihar, Vera Andreeva, Gregory and Laurie Wood, and Jessleen and Mini Kanwal for their interest and constant support that helped in getting this book finished.

To all of our pets and the wildlife around us that inspired this book. And to our parents who encouraged us to indulge in and pursue our interests.

Introduction

Where Are All the Dead Animals, Sri Lanka Asks

Wildlife officials are stunned – the worst tsunami in memory has killed around 22,000 people along the Indian Ocean island's coast, but they can't find any dead animals. Giant waves washed floodwaters up to 2 miles inland at Yala National Park in the ravaged southeast, Sri Lanka's biggest wildlife reserve and home to hundreds of wild elephants and several leopards. 'The strange thing is we haven't recorded any dead animals,' H.D. Ratnayake, deputy director of the National Wildlife Department, told Reuters Wednesday. 'No elephants are dead, not even a dead hare or rabbit,' he added. 'I think animals can sense disaster.'

Reuters, Sri Lanka (29 December 2004)

What did the animals know that humans didn't? What alarms were sent out that humans didn't 'hear'? As the Industrial Revolution and the development of urban centres moved us from nature and the land to the promise of science and technology, we detached ourselves from the understanding of the animal world. Twentieth-century scientific practices placed humans on a pedestal of superiority, further separating us from our natural roots and surroundings. Now, science is beginning to take us back to nature, providing a window into the minds of other species.

This book represents the coming together of two individuals with quite different backgrounds – a neuropsychologist and a neuroethologist – but with a common interest in the wonderful and secret lives of animals. Although the essence of this book was brewing in our psyches for most of our lives, it came as a compelling and timely surprise. Here, we provide a unique perspective on how to better understand the animal world and in so doing gain a better understanding of our own world – the inner world of our minds and the outer world that we share with all other creatures as our only home.

Bats sing, mice giggle

In a kapok tree growing in the tropical heat of a forest in Peru hangs a small male bat that has tiny sacs under each of his wings. Nine females surround him, each carrying a strong smell of a secretion that exudes from the sacs. The sac-winged bat feeds on tiny insects and interrupts its solitary existence to engage in reproductive activities. Intriguingly, pups of the species were discovered recently to babble. Four- to eight-week-old bat pups make long strings of barks, chatters and screeches that represent jumbled-up adult-like calls. Scientists now know that bats, like some primates and birds, babble as babies; and the ability to babble can even be accompanied by giggling. Not only do human infants babble and giggle as they experience feelings and try out their audiovocal abilities, so do babies of other species. New and sophisticated technology is taking our understanding into the secret world of animals where we can detect first-hand bats that do indeed sing and mice that actually giggle.

This book will take readers on a remarkable journey, during which they will discover that many of the behavioural and mental traits that have been considered to be uniquely human are in fact shared with other species. We'll show how animal 'friends' keep in touch, and how they warn and help each other in times of danger. We'll explain how some animals problem-solve, how they build and create, and how they entertain themselves and others. Some animals have a sense of humour; for example, parrots have been known to tell jokes of their own composition. We'll show too how parents of many different species, including bats, hug and cradle their young. And we'll also show how animals express grief and reverence in ways we never thought possible.

What did the animals know?

We set the stage in the first two chapters with cutting-edge findings that show how animal life depends on the strong electromagnetic fields circumnavigating our planet, and the weak electric fields and even weaker electrostatic radiations emanating from animals' bodies, as well as the vibrations they produce and detect.

All animals live in a milieu of electromagnetic waves and mechanical vibrations which they use for many of their transactions. For example, schools of electric fish generate complex electric fields and have sensors that can detect tiny distortions in these fields. Electric fish can use these electrosensors to find food, and to determine the precise location of prey or other electric fish when socialising. To solve the mysteries of how animals negotiate their surroundings and use their brains and nervous systems, scientists are delving more thoroughly into the realm of vibratory signals and even possible quantum-level occurrences that direct and surround all life.

Our sensory experiences define our lives and separate our world from that of our fellow creatures. Almost every species occupies a unique sensory niche in which it can find food and compete with others occupying the same niche. Some, like electric fish, become specialists in trying to overcome the competition. Others, like many species of bats, use sound pulses and their echoes to find food and probe their environment, in ways similar to those adopted by much larger creatures such as dolphins and whales in the deep and dark oceans. Yet others, like catfish, may retain an ability to live in diverse habitats as one of their senses becomes exquisitely developed.

While humans may never be able to experience the sensory niche of another species, by understanding more about the lives of different animals, we're able to learn useful tricks to aid our own survival. We can't, for example, imagine what a walk in the woods smells and feels like to a dog, which has such olfactory acuity. However, canines have long been used by humans to find missing people or sniff out illegal substances. Recent research has even shown that dogs can detect breast or lung cancer in a person by smelling that person's breath. Cancerous cells produce different metabolic waste products than normal cells, and dogs can smell that difference.

Exploring these sensory capacities further in Part II, we focus on the survival strategies that animals adopt when confronted with extreme environmental conditions, or with a threat from predators or their own kind. Despite their different sensory experiences, all animals, humans included, are endowed with a deep desire to survive. We will investigate alarm behaviours in a diverse range of species, from bees to cats to sharks, connecting the dots to attempt to answer the provocative question we first posed: What do animals know that humans don't know or heed when danger strikes?

Natural disasters are among the more dramatic threats that animals have to face. Day-to-day existence brings with it its own regular set of adverse conditions, from the fluxing and waning of food supplies to the annual cycles of seasonal weather. Less immediate than an earthquake, the winter months when life is at its lowest ebb may present the sternest of challenges to even the toughest of animals.

In Chapter 6, therefore, we discover how some animals employ hibernation behaviours, and that these can be adapted to combat extremes of heat as estivation. From the brown bear that gives birth during hibernation to the Antarctic cod that lives in the freezing waters of the Southern Ocean, we will look at what happens to animals physiologically and psychologically in these conditions. Not only that, some animals go into a light hibernative state every day to conserve energy – hummingbirds, for example, may conk out for several hours. Yet despite this, hibernation is very different from sleep and we'll study the odd way in which they may work together. By focusing on sleep processes in the animal world, we'll come across the many different approaches to it, including animals that go to sleep with only one half of their brains at a time. And as we draw together the latest scientific research on sleep and hibernation, we will endeavour to answer the question of not only whether animals dream, but what it is they might dream about.

Chapter 7 discusses the seasonal migration of animals as they move to warmer climates or fresh food sources. Some mass journeys, such as those made every year by thousands of wildebeest in the Serengeti, are breathtaking displays of beauty and power. Other trips are just as breathtaking, but for wholly different reasons. The distances and methods involved in the long flights of butterflies and birds, for instance, have stunned scientists. The tiny brain of the monarch butterfly can calculate distances and directions that would confound the most skilled airline pilot. And migrating birds in their hundreds fly non-stop for days, even weeks at a time, yet know to stop en masse if one of the flock is sick or exhausted. Such migrating and swarming behaviours in animals have become especially important research topics over the last few years, now that we have the technology and mathematical theories to help explain how animals know when and how to flock together and where to travel.

In the last part of the book, we explore communities of animals and the emotions and desires that they produce and that we all carry with us. Our connection with all other animals really drives home here. We not only inherit these basic desires, but all of our emotional expressions stem from them – to laugh, to play, to have sex, to reproduce, to deceive, even to kill: all derive from desires existing within animal minds.

New research is discovering that it's the deep limbic parts of our brain that secretly drive all of our thinking and our so-called 'rationality' – this is the centre of our emotions, of our quick evaluation and reaction against danger, of our memory; it regulates our system without the need for constant conscious analysis. The limbic brain is the primitive brain; it's present in all vertebrate species and probably has analogous regions in most invertebrate brains that we haven't yet discovered or even started to look for.

It's our large neocortex, though, comprising 30 per cent of the human brain and responsible for high-level thinking, as well as cognition and speech, that humans have contended sets us apart from other animals. However, recent studies in the wilds of New Guinea of the long-beaked echidna, one of the oddest and most enigmatic members of the animal world, have thrown a startling new light across this assumption. Belonging to the monotremes – a sub-set of mammals that lay leathery, reptile-like eggs, and which also includes the duck-billed platypus – this rarely observed creature has an electroreceptive, hairless tubular beak, webbed feet, spiny skin and, in the male's case, a bizarre four-headed penis. Most interestingly though, is that the neocortex of this peculiar animal is proportionally larger than that of a human, accounting for a remarkable 50 per cent of the brain. What is this spiny monotreme, the size of a terrier, doing with such a large part of its brain devoted to what is for humans the seat of analysis, language and consciousness? What can we learn from the echidna?

As we study the neurobiology, physiology, behaviour and individual experiences of animals, we have to be careful not to anthropomorphise. The fact that ants have cemeteries for their dead doesn't mean that they mourn in the same way we do. We can't even say that they mourn at all. Yet it's important to get to the truth. More and more studies are helping us do this. For example, a recent experiment has shown that dogs do indeed have a sense of fairness. They get jealous if they feel another dog is being treated better, and will withdraw affection or stop being cooperative. Researchers have provided examples of how animals protect each other and offer consolation to those that are upset. A variety of birds and mammals show empathy and altruistic behaviour. Grief-stricken apes are known to have carried their dead babies around with them for days. Elephants have been filmed caring for the sick, and slowly walking for hours and hours around the dead body of a member of the herd. There are countless stories of pets mourning the loss of a loved one. It's only recently that our technology has advanced and our research has become comprehensive enough to confirm some of the anecdotal evidence that has been around for so long. Many of the intuitions that humans have always had about our links to other animals are turning out to be correct.

★

As long as each generation continues to live closely with other animals and they with us, we will retain a modicum of sanity. Our pets and the wild birds around our houses accommodate to our ways and attempt to train us to theirs from the day we are born. Such interaction and involvement not only create awareness and modify behaviours, but have been found to change genetic information deep in the cells of us all. As we learn deeper truths about the life around us, perhaps we will also be able to better appreciate our own being and relationships, and even expand our capacity for affection, solace, hope, and love.

Part I

Sensing

1

A Supercharged World

Not only is the universe stranger than we imagine, it is stranger than we can imagine.
Sir Arthur Eddington (1882–1944)

Electric fish jam the frequencies of rapidly changing electric fields generated by their rivals. Birds 'see' the magnetic lines of the earth. Under the water, on land, and in the air there are electric and magnetic fields that affect all life. While most humans have little conscious awareness of these electromagnetic influences, medical science knows better. Physicians evaluate our health by measuring our heart and brain waves, which are also electric in nature (like those of electric fish, human brain waves change more than 100 times a second, as demonstrated by Stanford University's Karl Pribram). Modern medicine also uses the ability of our atoms to align to magnetic fields for MRI (magnetic resonance imaging) scans. This chapter explores how various types of animals sense and use electricity and magnetic fields to communicate with each other, to get around, and to protect themselves from danger.

The world that we, as humans, live in every day seems to us an immutable expression of how things are. We can draw it in through our senses, a world of reassuring 'objective' facts, full of colour and music, aromas, tastes and tactile surfaces. These five key senses inform our understanding of how and what the world actually is. But think a little further about the myriad of other species that inhabit the land and sea and air and occupy the same world as we do. Our assumptions of the physical space around us become difficult to translate into the experience of other creatures.

The world as we perceive and experience it depends entirely on the range of stimuli that our senses can detect and what we care to pay attention to. For example, visible light – to which the photo-pigments in our eyes are sensitive – is a very thin slice of the vast spectrum of electromagnetic radiation (from shortwaves and microwaves to X-rays and gamma-rays) present in our environment.

Rather than there being one objective world that is home to all living organisms, there are multiple subjective worlds. Nature believes in pluralistic thinking in an evolutionary sense, where different species occupy their own cognitive niches, each replete with unique angles and perspectives. Every species has its own way of perceiving and interacting with its environment based on its size, its surroundings, its sensory capacities, and how it behaves and what it remembers. The underwater experience of fish or the airborne moments of those birds and bats that fly can only be imagined by those of us who live on land. A bird's-eye view is one that we humans often envy, but we can't begin to imagine what the world looks like to a mantis shrimp that has many complex eyes that rotate in separate directions and are sensitive to the wide spectrum of wavelengths ranging from ultraviolet to infrared.

The senses that we are familiar with as humans are just the tip of a fantastic and fascinating sensory iceberg. It's below the surface where we enter a realm of senses more potent than those of most humans. There are electroreceptive animals for whom electric fields inform everything about the world in which they live, and there are other electrogenerative species which can not only pick up on such fields but generate their own electric charges. Completely beyond what we humans may consider the world to be, these electric fields influence everything from the way animals build and navigate with their own subjective 'maps', to the way they hunt and the way they interact with one another.

Scientists readily agree that many different animals, including millions of migrating birds, use the earth's magnetic fields for navigation and migration. Studies of homing pigeons suggest that they do genuinely 'home in' using their own bandwidth of information about spatial patterns in the earth's magnetic field.

However, we're only just beginning to understand the various ways in which this may occur. While we have studied groups of 'electric fish' for decades, we're still trying to figure out exactly how this sensory system functions and how it achieves the amazing feat of extracting information from transient distortions in electrical fields. Yet the pictures of these animal worlds being fed back from the forefront of scientific inquiry are remarkable. Sharks, skates and rays receive electrical information about the position of their prey, the drift of ocean currents and their magnetic compass heading. In training experiments, stringrays show the ability to orient themselves relative to uniform electric fields similar to those produced by ocean currents. Researchers have also discovered complex patterns of electrical discharges generated by electric fish used for 'flirting' with each other and making mating choices.*

Water is well suited for electrolocation in animals. While it takes a very high voltage for electricity to cross the air barrier, water, especially brackish and sea water, is highly conductive. Consequently, much electrical field research centres on animals that live in the water or spend a great deal of time in wet areas, because it's much easier for our present research instruments to pick up their electric signals.

In a way, therefore, humans and all other living things are permanently plugged in to a vast electromagnetic, biologically viable resource. Only certain animals, however, are able to use the electromagnetic fields directly to shape and define the worlds in which they live. While scientists at several US universities have taught human subjects to focus their thoughts to affect mechanical devices like computer cursors and prosthetics for paralysed individuals and those who have lost limbs – and it has been projected that in the not-so-far-off future, humans will be able to guide

---

* For example, Philine Feulner at the University of Sheffield and her colleagues found that female electric fish of the *Campylomormyrus compressirostris* species choose the males of their own species over males from very closely related species on the basis of the different electric signals they send out.

the flight of spacecraft with their minds – birds already travel thousands of miles without any mechanical devices or navigational equipment, and so do many seafaring creatures. These animals use electromagnetic information and direction to help them move, communicate and navigate over very long distances. Before we move on to looking at how they carry out this remarkable feat, we will start out by meeting two of the more peculiar electroreceptive animals around.

Animals that detect electric fields

Two unusual mammals are known to detect electrical signals. One looks like a furry beaver with reptilian-like flipper feet and huge flat bill – hence the name duck-billed platypus. It usually hunts after dark in murky river waters for crayfish and other crustacean delicacies, as well as worms or insect larvae. As the platypus sweeps its head from side to side, its huge snout, loaded with more than 40,000 electrosensors, sends signals to the brain to create a map of the electrical fields of its prey. The other electroreceptive mammal – the echidna – is peculiar-looking, with club-like feet and what resemble feathery porcupine quills stylishly covering much of its body. Its pointed beak, which comes in sizes long or short, is used for tracking down earthworms or other soil- or swamp-dwelling prey. The long-beaked echidna, mentioned in the Introduction, lives in wet tropical jungles and has around 2,000 electroreceptors in its beak that pick up the electrical signals of prey. The smaller, short-beaked echidna that lives in drier regions has more like 400 receptors at the tip of its beak. It usually waits until it rains before finding its meal – electrically, of course, in what is called electrolocation.

Whether searching for a good meal or a good mate, many animals that sense electricity around them use ampullary receptors. An ampulla is a pore in the skin that leads to a canal filled with gel. For example, the dark spots on a shark's snout are often sense organs called ampullae of Lorenzini, after the Italian zoologist Stefano Lorenzini who described them in 1678; it wasn't until the 20th century, however, that scientists understood their function. And a shark can sense electrical fields such as those given off by the muscle movements of another fish, because the ampullae read the difference in voltage between sensory cells at the pore and the base of the canal – the system basically functions as a voltmeter.

Fish that generate electric fields

For many of us, the only fish we associate with electricity are electric eels, and we assume that they would zap anybody who gets too close. However, electric eels are just one of many different species of fish capable of electrogenesis. The various types of these live-wire water-dwellers fall into two distinct categories – weakly electric and strongly electric fish. In both groups, electric transmission, as in visual systems, is nearly instantaneous and is little affected by 'noise' in a system, but doesn't go far and thus is effective only over short distances. Like chemical and sound communication, an electric signal can also pass around objects.

Within the two classes, some species are known as 'pulse fish' because their electric organ discharges, or

EODs as scientists like to call them, are sent out at a low and irregular repetition rate (at intervals several to many times the duration of a single pulse). Other fish species are 'wave fish' because the inter-pulse intervals are brief and regular (equal to or little longer than discharge duration), reaching regularities higher than any other known biological rhythm. Walter Heiligenberg, working at the Scripps Oceanographic Institute, discovered that electric fish can shift the frequency of the EOD if two fish producing the same frequency approach each other. Since the purpose of this shift is to avoid jamming of the two signals and loss of information, he called it the jamming avoidance response or the JAR.

Behavioural uses of EODs and electroreception by electric fish include: disorienting and confusing potential prey and potential predators, or finding prey (even if buried under sand); determining location (electrolocation and electro-orientation) by echo or by interaction with the earth's magnetic or electric field; social communication (including reproductive behaviour); and sensing of weather conditions, time of day, earthquakes, and distant lightning. And, as we have seen, even a system of avoiding being jammed by each other's signal has evolved in several species of fish.

Strongly electric fish produce much higher voltage pulses than their weakly electric cousins. The strongly electric fish category includes not only the electric eel, but the electric catfish and the electric or torpedo ray. Fish with electrogenesis usually produce electric charges through electrocytes or electroplaques – typically flat, disc-like modified muscle or nerve cells all stacked together. In marine fish, these are connected like batteries in a parallel circuit, whereas in freshwater fish these are connected like batteries in series. These latter are capable of producing discharges of higher voltage, necessary as fresh water doesn't conduct electricity as well as salt water. Each cell in the battery can produce nearly 0.15 volts by pumping out positive sodium and potassium ions. The electric eel* has around 5,000 or 6,000 electroplaques in its abdomen, allowing it to generate shocks as strong as 600 volts to stun its prey. It can also lower its voltage pulses to around 10 volts to navigate and to detect prey. As a result of its electrogenerative abilities, the electric eel is found only in freshwater habitats, as salt water can have the unhelpful effect of causing the fish to naturally short-circuit. The electric ray, which has no such problem, has a pair of special kidney-shaped organs at the base of the pectoral fins that generate and store electricity and send out charges from 8 to 220 volts to electrocute prey or to stun a possible predator. Electric catfish, common in African fresh waters, generate their electrical discharges from their skin rather than from electric organs that consist of individual electroplacques.

When a foraging torpedo ray detects prey it swims forward and upward, exposing its ventral surface towards the fish while emitting low-frequency voltage pulses. The currents passing through the victim's body excite its nerves and muscles, stunning it and immobilising it, whereupon the torpedo descends over it and consumes it while continuing to emit pulses. Large Atlantic torpedo rays can generate

* Technically the electric eel is not an eel at all, but a type of knifefish.

enough power to produce a shock of up to 220 volts; that's enough voltage to run your everyday appliances, such as a mixer or clothes dryer. Of course, animals can't deliver such voltages in a sustained way, as the purpose is simply to stun the prey. So it's more like carrying a stun gun inside your body. Smaller rays, like the lesser electric ray (*Narcine brasiliensis*) can only muster a shock of about 37 volts because their prey are smaller in size.

Weakly electric fish, like the elephantnose fish, use their electrogenerative ability either to navigate or locate prey, or to communicate. Instead of the electroplaques as in the electric eel, they have electric organs consisting of columns of electrocytes that generate relatively feeble electric fields. This type of active electrolocation relies on the ability of the elephantnose fish and other species that use it to detect any distortions in an electric field of less than 1 volt. Although it's not well understood how the brain extracts all of the sensory information for active electroreception, the sensors (also called electroreceptors) in the skin are sensitive to the rate of change of voltage across their cell membrane. In contrast, sharks and rays as well as most species of catfish use passive electroreception where the animal senses the weak bioelectric fields generated by other animals. Sharks are the most electrically sensitive animals known; they respond to DC (non-alternating current) fields as low as 5 nanovolts per centimetre and can use this ability to detect a small fish buried in the sand.

Sometimes elephantnose fish deliberately get into 'jamming' – not the jazz type, but similar in a way to when the musical session becomes a showy game of one-upmanship. Instead of competing or having fun with music, the electric fish try to jam the electrical signals given off by another fish. Sometimes they jam to get a mate they want, sometimes to get a meal. Ghost knifefish found in the rivers and streams of Central and South America are also known to jam the frequencies of other fish. Unlike other electric fish, in which the electric organs are derived from modified muscle tissue, in adult South American knifefish, the electric organ is a modification of the axons of spinal neurons or nerve cells. Interestingly, this transformation takes place only during the adult stage; as embryos they must rely on muscle-derived electric organs, which is therefore considered the more primitive system. Among groups of knifefish, the strongest adult male lets his presence be known by sending out the highest rate of electrical discharge – 900 pulses per second. The rest of the males have lower frequency rates and are careful not to interfere with the frequencies of others. However, when a rival wants to challenge the dominant male, the challenger will also generate pulses at a frequency of 900 hertz to jam the signal. In response, the dominant male might zap his rival with a higher frequency charge of up to 1,000 hertz. But if the rival male continues to challenge and jam, the two males will probably fight it out physically, sometimes ending up locked jaw-to-jaw for an entire night. So jammers beware.

Magnetic fields

While electric pulses and waves can be used as a tool for finding your way around a localised environment, for much longer journeys it's the ability to navigate using the earth's magnetic field that really counts. Unlike electric fields, magnetic fields are powerful force fields that are present globally. The earth's magnetic field is a geomagnetic dipole field where the poles lie in the vicinity of the geographic poles. Movement of the earth's fluid, outer iron core generates the magnetic fields, although details of this theory remain unclear. Despite reversals in polarity after several thousand years, as well as temporal and regional variations on a smaller scale, these fields can be considered to be fairly stable.

Migrating birds use these magnetic cues to fly hundreds or even thousands of miles on a round trip every year to escape harsh conditions and find food and safety. The pink-footed goose, for example, makes its way each winter from Iceland to the more temperate climate of Britain. Even more impressively, the bar-tailed godwit undertakes an incredible annual journey all the way from Alaska to New Zealand. Covering 7,258 miles (11,680 km), it's the longest non-stop flight of any bird. Many more bird species also make impressive migrations.*

Birds are not the only animals to navigate across huge distances in this manner. Insects like monarch butterflies, fish, turtles and mammals of the sea travel with amazing precision. Many animals are able to detect the earth's magnetic field and use it for navigation and other reasons – such as the mysterious reason behind why cows have been found often to stand in north–south positions or why termites align their giant mounds north to south. Several carefully constructed experiments confirm that many animal

---

* The sheer exhilaration and peril of vast bird migrations is captured on film in Jacques Perrin's magnificent *Winged Migration*.

species have internal compasses. Even the wispy monarch butterfly with its tiny brain smaller than the head of a pin has its own compass and timekeeper. In addition to the north–south compass reckoning ability, certain migratory species are thought to also use the geographic differences in the strength of the earth's magnetic field as navigational guides. Birds may even be able to see these fields.

As humans, we are beginning to see more with our instruments. And an international group of scientists, for example, has been working since 2003 on developing and updating a geomagnetic map of the earth. The earth's magnetic field is not uniform; it has many irregularities, such as the large one in the region of the Bermuda triangle where a number of ships and aircraft have gone missing.

Bird species such as homing pigeons and a number of types of turtles have been studied extensively to determine how they detect and orient themselves to these magnetic fields. Loggerheads and green sea turtles are favoured subjects. These sea turtles swim across an area of hundreds of kilometres and find their ways to the same nesting and feeding sites year after year. Genetic analyses have confirmed that the adults of at least some populations do indeed return to their birthplace for nesting after first migrating to distant oceanic areas. How do they do it? Ocean waves are useful guides in shallow ocean waters, especially for hatchlings that have no previous experiences. Although these cues disappear in deeper waters as the waves become refractory, the sea turtles are able to maintain their orientation. Experiments have shown that turtles can differentiate between varying magnetic field intensities found along their migratory route; such an ability is a prerequisite for using a magnetic map.

Research into finding the sensory mechanisms that enable receptivity to magnetic fields in more complex organisms has been slow because of the difficulties in designing clean experiments that generate reliable data. The phenomenon of magnetoreception, however, is now well established, and this ability is known to be present in several groups of invertebrates (molluscs, annelids and arthropods) and vertebrate species, including humans. A type of marine slug called *Tritonia*, a species known to use magnetic fields for orientation, holds one of the proofs of complex magnetoreception. Kenneth Lohmann and his colleagues at the University of North Carolina, Chapel Hill, recorded tiny electrical impulses in a neuron known as PE5, which subsequently increased its firing rate (the frequency at which impulses are generated) every time the scientists rotated the magnetic field by 20°, 60° and 90°. This provided clear evidence that a magnetic cue was being transduced or transformed into an electrical signal inside the nervous system. This meant that the information was now available for the nervous system and could be used to guide behaviour, although how exactly this is brought about and where the sensor is remains unknown.

While scientists aren't sure how animals perceive them, they think there are at least three different ways in which animals can orient themselves to the earth's magnetic fields. And scientists agree that different animals may use more than one method.

As children, many of us have played with magnets and small iron filings. In fact, the way those filings always turn around towards the magnet is similar to one way scientists think some animals respond to magnetic fields. One hypothesis is that something very similar to those little iron filings is present in the bodies of animals. Crystals of magnetite, a magnetic form of iron oxide, have been found in a number of species – including near the beak of pigeons, in the head of trout, and in the abdomen of honey bees. Clusters of these ferrimagnetic minerals seem to be affected by the intensity of the earth's magnetic field. This internal compass in its simplest form is seen in magnetotactic bacteria that have small particles of magnetite arranged in chains. They line up according to the earth's magnetic field, allowing them to be propelled northwards in the northern hemisphere. Interestingly, bacteria found in the southern hemisphere have their polarity reversed and move in a southerly direction.

Scientists' search for magnetite in animals is difficult because most, if not all, animals have iron distributed throughout their bodily tissues. This is certainly true for humans. In fact, iron is among the most common metals found in our organs, and in some degenerative processes such as haemochromatosis (where iron doesn't get broken down and absorbed into the body properly), Parkinson's disease and blood coagulation, iron can accumulate.

The second way in which scientists think animals may sense magnetic fields is called the 'radical pair model'. This model involves certain biochemical reactions, and some subatomic (quantum) mechanisms that change the spin relationship of two unpaired electrons.* It seems that photoexcitation, an energetic response to light, would start such a reaction, and scientists are very interested in some photoreceptive proteins known as cryptochromes. Cryptochromes are found in the right eyes of migrating birds. When light passes through a layer of cryptochromes, a series of exchanges results in what scientists call 'optical pumping' to generate lone molecules in an excited state – they vibrate faster and are out of stability. A pair of these radical molecules works together so that each ends up having an unpaired electron† in an excited state. This type of chemical reaction can be triggered by a single photon of light and the unpaired electrons that are generated are said to have a magnetic spin or 'moment'. At literally that moment, they can interact with a relatively weak magnetic field in various ways before switching to a more stable state. Thus the direction of the earth's magnetic field alters the spin state and this can be translated by the magnetoreceptive system of birds to yield navigational information. This makes the task of orienting to a geomagnetic direction relatively easy for the birds, and could result in weird illusions such as the horizon turning purple if a migratory bird such as the European robin looks in a certain direction.

* Electron spin, which is a principle borrowed from quantum mechanics, is a theoretical potential of an electron. If the finer details aren't immediately apparent, take solace in the fact that aspects of quantum physics bamboozle even the scientists who work with it.

† In the subatomic world governed by the laws of quantum mechanics, two unpaired electrons are electrons that are not chemically bonded. Individual electrons of this kind are very volatile and will react easily and spin at different angles and intensities – hence 'excited state'.

Electrophysiological recordings from the neural structures of the migrating bird believed to be involved in this process, the accessory optic system and the ophthalmic nerve, have confirmed the presence of a direction-specific electric signal triggered by magnetic fields in the presence of light. It's possible that avian navigation may be aided by both magnetite in the bill of a migrating bird to sense the intensity of the earth's magnetic field and to locate magnetic north, and cryptochrome in the right eye to sense the direction in which it's flying.

The third hypothesis is based on the phenomenon of induction, and would be most likely to occur in salty water conditions. According to this hypothesis, a fish swimming casually through the earth's magnetic fields causes the formation of an electric field.* Magnetic fields create an electric field due to the forward motion of the fish through the magnetic field. This electric field (or voltage gradient) causes a flow of charges in the electroreceptors that act as voltmeters and that were originally designed to detect electric fields. The faster the fish swims, the stronger the electric field, and it also varies according to the direction of the fish. The strongest electrical signal is obtained when fish swim east–west rather than north–south, since that way they cut across the maximum number of lines in the earth's magnetic field. All three hypotheses could be correct.

* This is obeying Faraday's law of induction which states that changing magnetic flux through an area produces an electric field on the boundary of the area. Faraday's insight was simply that it takes a *change* in the magnetic field to produce an electric current; the magnetic field itself is not enough.

For many years now, scientists have pondered the origins of life and its evolution here on earth. Life is supposed to have originated in a chemical soup by discharges of lightning, leading to the generation of new and complex molecules that could replicate themselves by a marvellous process. What we have to stop and consider now is that the chemical phenomenon that we call 'life' is highly sensitive to operations at atomic and even subatomic levels. At these levels, electric and magnetic fields can have literally invisible and unforeseen effects on the state of our molecules. Organisms as simple as bacteria evolving under these conditions developed to use the omnipresent magnetic fields to their advantage. Since all life evolved in the presence of electromagnetic forces and fields, it may not be so surprising to see that many animal species make use of electricity and magnetism in intriguing ways. Only recently have we begun to appreciate and understand how these natural fields are used by various species, and how some species have even evolved ways to generate their own electromagnetic fields as tools to orient, to navigate and even to communicate with others.

2

Good Vibrations

Nature does nothing uselessly.
Aristotle, *Politics* (350 BCE)

It's a steamy summer night. A solitary mosquito, sensing which way to fly for her next meal, is suddenly assaulted in mid-air by a dangerous vibration. Instantly, she drops down three feet in flight. That was close! A moustached bat, fifteen feet away, was also looking for food, using powerful sonar to do so. He just missed our quick-thinking mosquito.

The mosquito depends upon its ability to detect subtle airborne vibrations as an early warning system. It's not alone. All animals use mechanical vibrations for many of their transactions, and everywhere different communiqués are being sent out to various life forms. And what is truly remarkable about these vibrations is that in many cases they are more than just the instinctive reaction of a mosquito – they carry meaning. The rumbling internet of the thousands of these vibrations transmits information all across the animal kingdom, and although many humans may feel immune to its message, we have nevertheless managed to develop – and continue to develop – a whole range of skills based on techniques learnt from our animal counterparts.

Physicist Ovidu Lipan writes in the journal *Science*: 'We live in a sea of vibrations, detecting them through our senses and forming impressions of our surroundings by decoding information encrypted in these fluctuations.' This chapter will examine cutting-edge findings about the vibrations that flood the animal kingdom. The use to which different animals can put this vibratory information is astounding. For example, scientists have found that the organisation of vast groups of individuals such as ant and bee colonies is achieved through vibratory communication. And it's not just the insect world that deploys vibrations to such good effect. They are used too by much bigger creatures – elephants that communicate over huge distances by vibratory means, and, larger still, whales that turn the tranquil world under the waves into an enormous sea of sounds.

The highly sensitive instruments at the forefront of scientific discovery have also identified vibrations in animal worlds that humans rarely perceive. As a result, we'll venture underground and into the night where the ability to interpret different vibrations means that lack of visibility is no longer a handicap.

But first we'll alight with the utmost care upon a spider's web, where we can appreciate even the subtlest activity of a tiny spider as an example of vibratory meaning.

Vibrations in the undergrowth

Most spiders have very hairy legs. That seems obvious as far as a tarantula is concerned, but the human eye usually doesn't discern the fact that smaller spiders also need a shave. When a single spider hair detects a vibration, that hair is directly connected with the dendrite (web-like wispy fibre) of a neuron, and the message goes directly to the brain. It's an exceptionally finely tuned sense of touch that enables a spider in the centre of a web to know exactly what is going on in the outer limits of its home.

Delicate vibrations through the female orb spider's web herald the arrival of a hopeful suitor. She lets him know of her interest by the type of 'jiggle' she sends back to him. Some insects, like the male water strider, send out vibrations by tapping the water to attract mates. Other insects, including some bees, have sensor organs in their legs to pick up vibrations from all around. They may use these to look for mates, or pick up the vibrational cues from the dances and waggles of other bees explaining how to get to the best flowers. At the same time, the buzzing of a honey bee itself can set the army beetworm caterpillar a-flitter – sending it into a quick drop-off-the-leaf manoeuvre, or at least curtailing its leaf-munching to listen for danger. Like many other caterpillars, the army beetworm has sensitive hairs that pick up airborne vibrations and is especially on the lookout for preying wasps. It doesn't realise that this particular bee is harmless and just out looking for the perfect flower.

Green shield bugs are masters of vibrations. Although they are thought to have originated in Ethiopia, they are now found worldwide and are capable of engaging in precise, meaningful (to them) and complex behaviour, just by using vibration. So adept is their ability to send messages to each other by vibration that it's an integral part of their mating habits.

First the male sends out wafts of pheromones to attract females flying nearby. But an interested female doesn't know exactly where he is. So she lands on a leaf and generates a sophisticated pattern of vibrations, sending out a strong and precise message: an initial pulse followed by an intense burst over a narrow frequency range, another burst over a broader range, followed by a five-second pause. Then she goes through the pattern again. The vibrations travel along the leaf and plant stem and can go into the roots and reach other plants at a rate of 98 to 330 feet (25–90 metres) per second. The male listens with his feet and straddles along two stems, gauging the direction of the message. He responds with five pulses that encourage her to keep sending out signals until they find each other. If a female decides she's not really interested after all, she sends a low burst of vibrations, and the male just moves on, trying his luck elsewhere. There are always other females flying by.

Vibrations in the ground

By paying attention to the physical vibrations of these shield bugs, we can open up a vast, unceasing world of complex communication.

When Karen was a little girl and people still played records on record players, her father showed her an experiment that shares its basic physics with much of the following forms of vibration. With a pin stuck into the eraser of an unsharpened wooden pencil, they put a 78" record on the phonograph and started it spinning around. When Karen held the wooden end of the pencil between her teeth and carefully put the pin between the grooves of the spinning record, she was able to 'hear' the music playing on the record. Vibrations from the pin touching the record were being channelled straight through her teeth, up her jawbones to the bones in the ear, and she could hear the music very clearly via bone conduction.

Many creatures across the animal kingdom use substrate-borne signals (vibrations that travel through material substances like the ground, water, or bones – in contrast to airborne signals) to keep in touch with each other. This has been studied in all animals from insects to mammals. There are many different ways in which these animals send out vibrational signals, and many types of sensory organs that they use to detect them. Elephants, for example, can pick up seismic (ground) signals through their trunks and feet which are exquisitely sensitive to vibrations. A male elephant can pick up the low seismic rumble sent out by a female in heat over 6 miles (9.6 kilometres) away. The seismic vibration travels through the elephant's forelegs and up through its bones to the special bones in the inner ear.

Of course, elephants have an advantage over many other animals by simple dint of their size which enables them to create vibrations with the necessary strength to travel significant distances. For much smaller animals, it's those that live within the ground itself that benefit the most from seismic vibrations.

Beneath the ground, there's virtually no light and eyes are useless for purposes of exploring one's environment.

In order to find out what the rest of the blind mole rat community is up to, therefore, blind mole rats sit in their underground passageways, holding their jaws against the earth to hear messages being transmitted by another mole rat which is literally hitting its head against the wall of its own isolated tunnel.

The blind mole rat is a model animal for studying intraspecific (within a species) seismic communication, because it lives its whole life underground. Rodents that live underground, and especially those that are blind and spend their whole lives under the earth, comprise a unique population to study because seismic vibrations can create very effective long-distance communication.

The blind mole rat's ability to hear airborne sounds is limited. Its vestigial eyes, covered by skin and fur, show some sensitivity to light, but basically it's blind and it lives in voluntary, comfortable solitary confinement in its tunnel system, which is dug to its own body width, for most of its life. It gets together with other mole rats only while it's a pup, while raising pups of its own, or when it's mating. The blind mole rat might also encounter another mole rat that by choice or chance comes into its tunnel. Even if the visit is accidental, it usually leads to serious fighting and the death of one or both mole rats. So to avoid unwanted encounters and yet enable males and females to get together for mating, an intricate seismic signalling system is established between neighbouring tunnels.

Scientists are learning about these elaborate communication systems by planting a geophone into the soil close to the mounds of the mole rats. By studying the vibratory bursts picked up by the geophone, researchers have been able to distinguish the individual mole rats and their tunnels. Just like the different brain wave patterns that have been found to be unique to each person, each mole rat has its own seismic vibratory signature.

The life of a mole rat starts in an underground nest chamber with its mother and the rest of the litter, where most of the communication is through airborne vocalisations. Then at about seven weeks of age, the young pups get tunnel-digging practice by creating their own initial tunnels as extensions of the maternal tunnelling system. At this point, the connection between the established and the new tunnels is still open and allows for continued vocal communication between the mother and pups. This lasts for a four-week period as the youngsters also start learning and practising seismic vibratory signals. Then at about twelve weeks the young mole rat ventures off on its own – sealing its connection to the maternal tunnel and digging its own discrete tunnel system where it will spend almost all of its life. Now the communication is through the seismic channels. And a great deal of 'talking' goes on between these lone, isolated mole rats; a lot of head-drumming and jaw-listening. Scientists have picked up vibratory 'dialogues' between individual mole rats as far as 15 metres apart. This type of tuned awareness to tremors in the ground corroborates the anecdotal evidence of the acute sensitivity of animals to earthquakes.

How does this 'mole telephone' work? One mole rat bangs against the roof of its tunnel with the flattened, bony, antedorsal top of its head. The receiver presses its little cheek and lower jaw against the tunnel wall to get the latest vibratory message. Researchers are looking at two sensory systems that may be involved in this transaction. A few scientists suggest that there may be one 'receiver set' in the form of the animal's somatosensory system (a special area in the brain) through mechanoreceptors (sensory receptors that respond to mechanical disturbance). However, the most prominent – and thought by some to be the only – sensory system involved is the auditory system that receives messages by means of bone conduction (similar to Karen's 'pencil-between-the-teeth' hearing). The message is transmitted by the jawbone to the inner ear.

A relative, the Cape mole rat, doesn't bang its head on the tunnel ceiling. Like the elephants and many other above-ground species including rabbits, the Cape mole rat thumps its feet on the floor of its burrow to get its message across. When looking for a mate, each mole rat will drum its feet at a special tempo to signal its sex. When the 'right' male and female locate each other they will interact at a distance by drumming together in synchrony. Not only does it find a mate by drumming, the Cape mole rat alerts possible prey such as earthworms that it's nearby, which is clearly something of a disdvantage of the system.

And like many other animals that are aware of seismic underground vibrations, all types of mole rats are acutely in tune with natural earthly rumblings, especially the Rayleigh and Love surface seismic waves that can be harbingers of earthquakes, as we will see in Chapter 5. These animals not only use substrate-borne vibrational signals to communicate with each other, they resonate to the natural vibrations around them.

Resonance and synchronisation

In 1898 eccentric scientist and inventor Nikola Tesla unwittingly caused an earthquake in Manhattan while testing a tiny electromechanical oscillator. Innocently, Tesla attached the oscillator to an iron pillar that went down through the centre of his building on East Houston Street to the sandy floor of the basement. When he turned on the oscillator at certain frequencies, specific objects in the room started to jiggle. When he changed the frequency, things in another part of the room started moving.

Unfortunately, Tesla hadn't considered that the iron column to which the tiny oscillator was attached ran down to the earth beneath the building and his vibrations were being transmitted all over Manhattan. An actual earthquake started rumbling in his part of town. Outside, people were panicking and two officers were dispatched from police headquarters to check on this well-known mad inventor. Tesla, 'unaware of the shambles occurring all around his building, had just begun to sense an ominous vibration in the floor and walls. Knowing that he must quickly put a stop to it, he seized a sledgehammer and smashed the little oscillator in a single blow … With perfect timing the two policemen rushed through the door …'

This is an example of what is known in physics as 'resonance'. When certain vibratory frequencies cause other entities – from tiny particles to objects – to move, or to somehow lock into ('entrain with') the same frequency, and the molecules and atoms start moving in unison, they are resonating. Electrons in a circuit can oscillate in resonance. A more dramatic example is when an opera singer hits a loud, high note and a champagne glass across the room breaks into pieces from the airborne oscillations.

Vibrations resonate. Most physics students have played with tuning forks. When we take two tuning forks that are similar and strike one of them (A), the second tuning fork (B) will start to vibrate as well, resonating to A's frequency. But because of environmental interference, or perhaps because B has a speck of rust on it, the frequency that it transmits back to A might be a tiny bit different, and A will adjust accordingly so that their shared frequency is exactly the same – the two frequencies will synchronise. Synchronisation is the result of two or more sources giving off vibrational frequencies that entrain – lock into each other.

Synchronisation is an acoustic effect first written about in the 17th century by Dutch scientist Christiaan Huygens, who discovered the phenomenon when he was trying to create high-precision clocks – their ticking entrained. Such entrainments of clocks or moving pendulums or the divergent tones of organ pipes are common.

There are also many natural examples of synchronisation in animals. Crickets can chirp in unison, so can cicadas, as their sound oscillations become entrained. Frogs croak or 'peep' together. Fireflies synchronise their individual glowing bodies into a flashing, pulsating light show. Humans who spend a great deal of time together have been found to synchronise physiological functions as well as their moods and memory. For example, it's well known that women who live or work together tend to synchronise their menstrual cycles. And synchronisation often occurs in an auditorium when an audience that starts applauding in a chaotic way ends up with synchronised clapping.

Synchronisation and swarming behaviour

Mole rats are highly adept at using vibrations beneath the ground, and there are certain creatures that are equally skilled at sensing and deploying them under water. It's a sensitivity to vibrations that enables those densely-packed schools of little fishes to dart so quickly, changing directions in near-perfect unison. Fish and larval amphibians have what is called a lateral line composed of receptors (neuromasts) that detect the movements and currents in the water. The lateral line system runs down each side of a fish. In many bony fish, as well as sharks and rays, these neuromast receptors are located in a canal below the surface of the skin. This system makes the fish acutely responsive to any movements in the water. Their lateral line system allows these synchronised swimmers to sense the water currents and each other's movements so they can make it difficult for a predator to zero in on one particular individual.

In their comprehensive book, *Self-Organization in Biological Systems*, Scott Camazine and others summarise how synchronisation, patterns, complexity and self-organisation are intimately entwined in such phenomena as synchronised flashing of fireflies, swarming locusts in patterns, and flocks of thousands of birds flying in exquisite unison. A self-organising pattern at a global level emerges from surprisingly simple events and the behaviour of individuals responding to local information. In the March 2009 issue of *Science*, researchers verify the general principles of self-organisation in their study of oceanic herring.

Nicholas Makris and fellow scientists were able to study in great detail, and for the first time, how vast shoals of fish form and are directed in their movements. Although many species of oceanic fish band together in enormous shoals that can span tens of kilometres and involve millions of fish, researchers until now have only been able to sample these phenomena in their slow-moving vessels. Makris and his colleagues studied shoaling behaviour of Atlantic herring with Ocean Acoustic Waveguide Remote Sensing (OAWRS) – new technology that's capable of instant imaging and constant monitoring of fish populations over tens of thousands of square kilometres. The scientists confirmed that shoal formation was based on initial conditions and 'ensued rapidly when these conditions were satisfied'. As they report:

> First, we found that the pre-existing population density of diffusely scattered individuals had to reach a critical density of 0.2 fish per square meter. Given this, we found shoal formation to consistently commence in a highly organized fashion near sunset, apparently triggered by reduction in light level.

More and more fish came in, with specific leaders gathering groups at a few discrete locations.

> The emergence of leading clusters of high population density set off chain reactions that caused rapid growth into vast shoals. We found the growth to propagate

> horizontally outward as convergence waves emanating from the cluster initiation points, which appeared to act as sources of the wave action … Once vast shoals formed, they migrated at speeds consistent with the synchronous swimming of hundreds of millions of individual fish … toward southern spawning grounds on Georges Bank, apparently for synchronized reproductive activities.

The researchers demonstrated that: '(i) a rapid transition from disorder to highly synchronized behavior occurs as population density reaches a critical value; (ii) organized group migration occurs after this transition, and (iii) small sets of leaders significantly influence the actions of much larger groups.'

Unfortunately for freshwater fish that swarm together, they aren't the only aquatic inhabitants with a finely honed understanding of water-borne vibrations. One possible predator, the alligator, may be lurking in the swamps and rivers with its own type of aquatic motion sensor. Dome pressure receptors, thousands of little black dots spread around its face, especially in the jaw area, allow the alligator to pick up movements from other alligators as well as from prey that might be in the area.

Interestingly, these receptors were discovered only a few years ago by a graduate student working on her doctoral thesis at the University of Maryland. She found that these receptors are 'innervated' (activated) by the trigeminal nerve. The location of these receptors at the margin of the upper jaw explains why alligators float in the water with just their upper jaws breaking the surface. Placement of the receptors here maximises the chances of detecting the slightest splash in the water, which is the strongest stimulus for the alligator to open its mouth and devour its prey. Although not yet extensively researched, it's highly likely that organisation of these receptors all around the jaw also helps the alligator to locate precisely where the splash came from.

The incredible findings relating to the immense shoals of fish that populate the world's oceans can be generalised to group activity of all species, including flocks of migrating birds that 'decide' to land for a while because one bird is too exhausted to fly any further, and nests of ants that change their foraging habits en masse once the food supply changes.

Self-organisation on this level is behind any number of aspects of animals' lives. The giant termite structures in Africa and Australia are stupendous examples of self-organised building. When Karen spent time in the southern regions of Africa, she would watch and marvel at how the termites would build their red earth homes, as high as 20 feet (6 metres) – intricate edifices that from afar looked like grand statues. Bert Holldobler and E.O. Wilson discuss in *The Super-Organism* how self-organising behaviour is the impetus behind the nest architecture of super-organisms like social bees, wasps, termites and ants:

> The study of nest architecture is … a true exploration of a hidden world that holds unsuspected beauty, pattern, and complexity … Some social insects can build complex structures complete with air-conditioning and fortified 'castles'. But unlike human construction, there is no architect, no blueprint, no global design that

governs the course of construction. Instead, nest structure emerges through the self-organization of multiple workers interacting with each other and with the environment as they modify it.

Human vibrations

Not only are humans informed by vibrations in all of our senses – including mechanical vibrations that we hear and the subtle vibrations that our skin receives in order to feel something – but the versatility of vibrations produced by animals has long been apparent to their human neighbours. Anthropologists coined the term, the 'tom-tom effect', referring to communication methods that Native Americans borrowed from the wildlife around them. Whether it's a rock musician playing the drums or a giant woodpecker pounding on a hollowed tree, such airborne sounds can carry a long distance. Native Americans developed tom-tom drums to send their own messages. They were also known to beat out codes to each other via the same hollow trees that woodpeckers used.

Even today we're still trying to replicate some of the abilities we see in the animal kingdom, and new technologies are increasingly directed towards extending our sensory world to that of other creatures, so that one day, perhaps, we can sense an approaching tsunami or imminent earthquake. This endeavour is already well under way with the development of solid-state sensors (with millions of transistors on a silicon chip) for our aeroplanes, bridges and roads, biosensors in our clothes, and even nanosensors inside our bodies – all inspired by the amazing senses of the animals that surround us.

Levels of interaction

Our technology and advances in engineering and physics are taking biology and our understanding of life forces to another realm – the realm of interactive forces at all levels, from subatomic entities to a whole person or animal, to groups, to an entire cosmology. It's not only probable that self-organisation entails all of these scales, but that self-organising processes could not occur without involvement at all levels.

Advanced technologies allow those studying life processes to shift from the 1990s focus on cells and molecules to learning more about waves, oscillations, frequencies and subatomic (quantum) activities and their interactions with the molecular structures of life. In medicine, for example, such interaction is seen in an everyday medical procedure popularly known as an MRI – when used to measure brain function via changes in blood oxygenation levels it's known as fMRI or functional MRI.

Two examples of our study into quantum-level interaction in natural processes are the radical pair model discussed in the last chapter and a fascinating phenomenon known as entanglement – which is already showing great promise in cryptology, in this case being able to send a code that cannot be intercepted. At the end of the 20th century, physicists started with using photons (tiny packets of energy) to demonstrate entanglement. Under certain circumstances, when two photons have a close relationship, they become entangled. And (although it sounds like science fiction) if photon B travels thousands of miles away and photon A with which it has become entangled is changed in any manner, photon B will instantaneously incorporate the same change. So for example, a code can be sent out from A to B without any information occurring between A and B, thereby preventing interception.

According to J.G. Rarity of Bristol University, scientists are also working towards the use of satellites to distribute entangled photon pairs, so providing a unique solution for long-distance quantum communication networks. The early work on photons has progressed to demonstrating entanglement in electrons and neutrons, as well as atoms. And amazingly, scientists at NIST, the US National Institute of Standards and Technology, reported in the June 2009 issue of *Nature* that they are able to entangle the mechanical motion of two sets of vibrating ions. This leads to very interesting speculation about the involvement of entanglement in natural processes.

So from the smallest subatomic scale to macro-levels of vibrations and signalling, every part of nature is constantly in motion. All is interrelated and changing.

3

Sounds for Tracking and Talking

Not content with his experiments of plugging bat ears with candle wax and seeing them crash into large objects, Spallanzani attached various sorts of hoods and bags to the heads of bats, some covering one portion of the head and some another. He found that serious impairment of obstacle avoidance occurred only when the covering lay directly over the ears or the mouth, but that rather large pieces of material attached to the other parts of the head gave the bats very little trouble. He seems to have thought that the sound of the bat's wings or body as it moved through the air was reflected from obstacles and stimulated the ear. But in 1794, this hypothesis could not muster much convincing support, even in the mind of its author. Spallanzani's final views on the matter at the time of his death in 1799 are stated by Senebier to have been the following: '… the ear of the bat serves more efficiently for seeing or at least for measuring, than do its eyes, for a blinded animal hurtles against all obstacles only when its ears are covered.'

Donald Griffin, *Listening in the Dark* (1986)

In this chapter we turn our attention to sound waves and the ability of some animals to live in a world created almost entirely by sound. This biological sonar* was discovered by Donald Griffin in the 1930s. Young Donald was an undergraduate student at Harvard College in Cambridge, Massachusetts and was actively engaged in banding bats to study their migrations (banding is like ringing of birds, except that a band is placed on the forearm instead of a ring on the leg). At the time, he was familiar with the view that bats felt the proximity of obstacles with their wings. Meanwhile, in the Harvard Physics Department, Professor G.W. Pierce had virtually the only apparatus in existence for the detection and generation of a wide range of sounds lying above the human hearing range. In the winter of 1938, Donald approached Professor Pierce to ask him to use his apparatus, called a sonic detector, to listen to his bats. When a cage full of bats was held in front of the parabolic horn of the sonic detector, it seemed to come alive with a medley of raucous noises emitted from the loudspeaker. They discovered that the sonic detector was picking up sounds that they could not hear, in addition to the ones that they could. Releasing the bats in the room, they found that they could pick up sounds only when the bats were near the horn of the detector, but not when they were flying freely at other locations in the room. They concluded that the bats could produce a type of very high-frequency call note beyond the range of human hearing. It took another year for Donald, now working with a fellow student, Robert Galambos, to discover that bats indeed always made these high-frequency sounds as they flew around the room, but the instrumentation could detect them only when the bat

* Sonar: sound waves for navigation and ranging.

was facing the horn. Griffin and Galambos then repeated some of Spallanzani's experiments conducted well over a century previously, aware that Hamilton Hartridge, an English physiologist who had studied underwater sound signalling during the First World War, had suggested that bats might use sounds of high frequency and short wavelength, although he had not conducted any tests himself. With carefully designed experiments that required a bat to fly through a mesh of vertical wires, they determined that high-frequency sounds were indeed crucial for obstacle avoidance in bats.

This exceptional ability was later termed 'echolocation' – and, unlike electrolocation, echolocation happens in air as well as in water. Bats have mastered the use of sounds to navigate, survive and thrive better than any other species, including humans, though sound waves are also used to a stunning degree of complexity by dolphins and whales, as we'll see a bit later.

Most people are fascinated by bats. They are exceptional creatures in all sorts of ways: from their ability to echolocate, to their status as the only mammal that can achieve true flight with the almost soundless flap of their wings which is quieter even than a barn owl's. Bats come in many sizes, shapes and forms. Although they are rarely seen, at least a fifth of all mammalian species are bats. Their facial features are bizarre and may resemble a cat, a dog, an owl or a common rodent. But bats are not rodents – they belong to a completely different group of mammals known as the Chiroptera. Having evolved over millions of years, the eighteen known families of bats have occupied hundreds of ecological niches. Found anywhere from open trees to buildings to dark, enclosed caves, their elusive nature is due to the fact that most bats fly out of their roosts at dusk and are active in the middle of the night.

There's something magical about this flight of bats. Blind as they tend to be, bats never crash into a person unless they get trapped in a small room and are completely disoriented. Bats have some unique and unconventional aerodynamic abilities, and their flight, which is only beginning to be studied, may yet lead to the fulfilment of the human dream to fly independently and silently – perhaps with a device that flaps its bat-like wings and allows you to take off from the edge of a skyscraper. Unlike birds, bats don't have feathers – as they are mammals like us, they give birth to live young and suckle them. Their wings are like the webbed feet of ducks, except that they have very delicate membranes stretched between the digits of their hands instead of the toes of their feet. The legs too are bridged together with a tight yet flexible skirt or interfemoral membrane that is ribbed along the middle by a very thin tail. The tail can be fairly long and flexible in some species. This structure works perfectly in guiding and breaking a bat's flight – as it needs to, just before landing on a branch or the rough surface of a rock in a cave. The feet are delicate and flexible with long digits like the fingers of the extra-terrestrial alien in Steven Spielberg's classic movie, *ET*.

Bats can be classified into two groups on the basis of size alone. The relatively big bats (megachiropterans) are fruit-eaters and mainly belong to one of the more than 180 species of Old World bats that inhabit a huge stretch of the globe from northern Africa through south-east Asia and down into Australia. For flying animals without all the special adaptations found in the bodies of birds, the very largest species, such as the Australian flying fox, require a wingspan of more than a metre in diameter in order to lift their bodyweight.

However, it's primarily the small, or microchiropteran, New World bat species that use their wonderful ability of echolocation to catch tiny insects such as mosquitoes and moths in mid-flight. Not every bat echolocates, though, and some of the small bats have gone back to eating fruits, while others have adapted to lap the warm blood of other animals. There are, in fact, three species of vampire bats that live in Central and South America and feed on large birds, cows, horses and pigs (contrary to popular opinion, vampire bats will very rarely seek to feed on human blood). Common vampire bats are quite small and have strong legs, unlike most species of bats, which if asked to run can do little more than flop around like fish out of water.

The smallest of all mammalian species is a bat the size of a child's thumb. Found in western Thailand in Sai Yok National Park in the Kanchanaburi province, these bats usually reside in small holes or crevices formed by stalactites in caves. They go out at dusk – flying above the bamboo and teak to feed on insects that may be in the trees or on the wing in the night air. Known as bumblebee bats, these tiny mammals were first discovered in 1973 by Kitti Thonglongya and are therefore also called Kitti's hog-nosed bats. Just 30–40 mm in length, they weigh approximately 2 grams (about the weight of a penny). This sadly endangered species is under increasing threat from deforestation by the teak-logging industry.

How does biosonar work?

While we may be largely at home with the notion that many bats find their way around their environment by sonar, it's more difficult to really appreciate this singular achievement. Trying to explain to a group of scouts what it meant to be 'blind as a bat' and how bats living in total darkness could literally 'see with their ears', Jag found himself struggling to visualise the concept. Calling his son from the audience to help him demonstrate, Jag blindfolded him with a strip of black cloth, raised a finger in front of his son's face and asked: 'How many?' The reply came: 'I don't know, I can't see.' Jag then held a large metal tray vertically and directly in front of his son's face. Instructing him to yell loudly and then listen carefully, Jag either held the tray close to or away from his son's face and asked him to tell the audience whether it was in front of his face or not. Each time his son's answer was correct!

In fact, we all have a rudimentary ability to extract information from the quality of sound in an echo and know something about the objects in front of us, even if it's simply their presence. If you're yelling in a vacant room, for instance, it's relatively easy to tell the difference in the echo when you're facing towards the wall and when you're turned away from it, even with closed eyes. We know the difference almost innately. Humans who are congenitally blind have a heightened sensitivity to sounds and sometimes to smells. There is documentation of one child who lost both of his eyes in early childhood and learned to 'see' by making clicking sounds and listening for their echoes. Not only could he navigate, he could also have pillow fights and play video games with his friends by simply tracking the sounds accompanying the actions in the game. He used clicks to locate objects and events in his environment and avoid any obstacles in his path without touching them.

An extreme adaptation of the sense of hearing or audition, echolocation is the ability to locate objects and move around successfully using sounds. Bats living in the caves of South America stream out by the millions precisely at the same time every evening like dense wisps of smoke from the chimney of an old house on a cold winter day. They are going out for dinner, to catch insects and sometimes even fish – and they have to do it all using sound. In echolocation, pulses of sound are sent out, then received as an echo by the sender after reflecting off a physical object. Like aeroplanes flying by their radar* in stormy weather, bats can fly by their sonar in the pitch dark.

Perhaps surprisingly, bat sonar is limited to a few feet in range. Should you ever find yourself in the path of a looming bat, it will fly right at you almost silently and then swerve to avoid you at the last moment. The wide-open mouth proudly displays the sharp white teeth, but not even the slightest whisk of the wingtip is felt by any part of your body. In reality, of course, the bat is screaming its head off at 120 dB SPL (sound pressure level in decibels), which accounts for its wide-open mouth. We can't hear a peep of this because its emissions are at ultrasonic frequencies that

* Radio detection and ranging. 'Radar' is an object-detection system that uses electromagnetic waves to identify the range, altitude, direction, or speed of both moving and fixed objects such as aircraft, ships and motor vehicles. The term RADAR was coined in 1941 and used extensively during the Second World War.

are far higher than our audible frequency range. Were we able to, the intensity of the sounds would rival those of a revved-up engine of an aircraft about to take off. The bat is simply trying its best to obtain as much information as possible from the returning echoes of its screams in order to sense what might be in front of it. And amazingly, to the relief of the observer, it succeeds in doing this each time.

Although silent to us, bats, of course, are constantly tuned to the sounds they produce. The ears of the insectivorous bats are usually extraordinarily large, allowing them to collect and amplify very soft echoes coming from a particular direction. They can flick their ears at such high speeds that even a slowed-down playback of a video appears fuzzy. The alternate flicking of the ears, combined with aiming them in a particular direction, makes dog ears with the sharpest of hearing look quite clumsy. What the flicking, together with head turns, accomplishes is the localisation and characterisation of a tiny pulse of sound bouncing back from the wings and body of an equally tiny insect. Together with the ear, the whole brain of a bat is tuned sharply to the frequency or pitch at which it's emitting the echolocation sound pulse. The ear canal and cochlea of most bats are specially designed to amplify and increase the signal-to-noise ratio of the weak echo, and the brain is wired up from millions of years of evolution to tap into the structure of this weak echo and extract all that a bat needs to know about the insect. What direction is it flying in, what speed is it flying at, and where is it exactly located in space – or more importantly, where will it be located in space two seconds later when the bat is ready to capture it? Listening to the echoes of tiny bits of sound pulses is all it takes to answer all these questions in just a few hundred milliseconds. How the brain is wired to do so and how it actually accomplishes this massive computation in a very short time have been the focus of intense research by a number of scientists the world over, yielding some amazing findings and solving many of the mysteries surrounding echolocation by bats.

Scientists have discovered that the brain of bats has computational circuits, more advanced than the most powerful supercomputer ever built, that extract all of the relevant information in a sound signal almost instantly. All it takes is one short sequence of pulses. This is enough to calculate the distance of a target to the last micron and time to the last microsecond. Bats have to do this in order to succeed in catching thousands of mosquitoes each night until their bellies are filled up, emptied and then filled up again multiple times, to rapidly generate the energy to carry out their daily activities at high speeds and very high rates of metabolism. In order to support this phenomenal workrate, a bat's heart has to fill up with and squeeze out the blood over 600 times per minute, about ten times faster than the human heart. Place a baby stethoscope on a bat and it sounds like a motorcycle going down the street in third gear. And some bat species are known to sustain this rate of heartbeat for up to 34 years. This would be the number of beats we would need to live for well over 300 years – without ever having to go to the hospital or see a specialist in cardiovascular surgery.

Bats in the Caribbean

The night was long and dark until finally the moon came out from behind the clouds. The peeping of the frogs became even louder. The coqui frogs joined in the chorus. Fireflies were everywhere and the occasional buzz of mosquitoes was a reminder that they too were still out there. Jag and his colleagues were on a trip to snare bats and take them on a long journey to the United States, where their communication and social behaviour could be studied carefully in the semi-natural environment of an academic institution. Finally, they heard the soft sound of flapping wings – just what they were waiting for. The moustached bats were coming back from their feeding flights. The mist nets were placed at the small cave entrance. Of course, most insectivorous bats can detect the fine strands of a mist net using their sonar if they care to. But they didn't care this night. On a full belly, they preferred to glide back into the cave having feasted on the mosquitoes and moths that were plentiful in the hot and humid climate of Jamaica. It wasn't long before a moustached bat was safely placed inside a specially constructed dome-shaped cage.

The echolocation hunting skills employed by these bats to catch their mosquito banquet depends on the complex use of sonar. Moustached bats emit an ultrasonic pulse that has two components in it – an unchanging, constant frequency or CF component followed by a frequency modulated or FM component. For this reason, moustached bats are called CF-FM bats. If slowed down to lower its frequency and come within the range of human hearing, it sounds like 'Eeeeeeeeeuuu'. Several other species, such as the little and big brown bats commonly found in Europe and North America, emit only an FM pulse and are categorised as FM bats. A few species emit only a CF pulse and are called the CF bats. Typically, the CF and/or FM components each have multiple harmonics. Almost any naturally produced sound has multiple harmonics, and moustached bats make full use of this phenomenon. The foraging behavior of CF-FM bats is typically divided into three phases: the search phase, the tracking phase, and the capture phase. The hundreds of pulses across the three phases sound like: 'Eeeeeeeeuuu, Eeeeeeeeuuu, Eeeeeeeeuuu – Eeeeuuu Eeeeuuu Eeeeuuu – Euuu Euuu Euuu Euuu'. By turning their heads and shaping their mouths and lips, bats can focus the entire energy within a narrow sound beam directed at a target of interest as they modify their flight path to lock on and intercept the prey. They are rather like a super-precision helicopter directing its light beam at a particular spot as it flies around in circles.

As a bat approaches its target, the duration of CF is successively shortened and the rate of emission increased until it sounds more like a buzz, also called the terminal buzz. It usually terminates with the demise of the insect that it's tracking, which is most likely stunned by the barrage of very high-intensity sound pulses. Some – such as a mosquito, a grasshopper, a praying mantis or a nocturnal butterfly – which can hear a bat's high-frequency sound pulses, may evade capture by performing an acrobatic dropping manoeuvre, only to fall into the sonar beam of a second bat hunting close by.*

Bat communication

Different species of bats can modify echolocation sounds to use them in a communicative context, as in mother–infant interactions, and as we shall see in Chapter 9, they may even eavesdrop on foraging and echolocating individuals to gather information on a food source. Relatively little is known about bats' ability to communicate with each other. The brain areas and mechanisms that process these sounds and coordinate all of their social interactions are just beginning to be discovered. These findings, however, have the potential to help us understand our own ability to communicate using sounds – one of the hallmarks of human evolution.

The moustached bat that Jag worked with in Jamaica is found across the steamy climate of the Caribbean and Central and South America and is one species whose communication ability has been extensively studied. Like *Desmodus*, a genus of vampire bats, moustached bats are highly social and live in large colonies of thousands to possibly millions of individuals in a single cave. One wonders what they may be saying to each other. Studies by scientists

* To hear this sound sequence, all you have to do is walk next to a small lake or river on a fine summer evening and bring a bat detector with you. This little and relatively inexpensive device lowers the high frequencies into the human hearing range, and you can hear the bats hunting before you can see them. These detectors are used for many scientific studies of bats and even to identify a species, just like birdwatchers do by the song or call of birds.

have shown that moustached bats have an incredible repertoire of communication sounds. Clearly they have a lot to say, even though we may not know the meanings of all their 'words'. What is of particular interest to researchers is that moustached bats have a vocabulary of at least nineteen simple sounds that have been recorded in captivity, and there may be additional sounds that are not emitted in a semi-natural and restrictive captive environment. What this means is that moustached bats have a potential for a vocabulary that at a phonetic level is comparable to ours. Although the number of combinations of these sounds that they can make is severely limited to less than twenty, it has nevertheless been shown that bats can combine their simple 'syllables' into more complex sounds, not unlike the phonetic syntax we see in all human languages.

On the North American continent live Mexican free-tailed bats, which migrate north to Texas and Arizona. Their communication sounds have been studied over a number of years and are gaining increasing attention from neurobiologists who want to understand how these sounds are represented in the brain of these animals. Since these and other species, such as the moustached bats, are so advanced in their ability to communicate using sounds, scientists are hoping that they can teach us something about how speech sounds are represented in the brain of humans. It may seem like a far cry from bats to humans, but several breakthroughs have already been made on this front, not only from studies of communication but also from studies of echolocation in bats. In fact, much of our understanding of how the human brain processes complex sounds, one of the most important advancements in humankind, comes from studies of echolocation and call processing in bats.

Animals with acute visual abilities have a large 'fovea' on their retina where a high density of photoreceptors can scan every little detail of an image. This is what allows an eagle to see a mouse running in a field from miles up in the sky on a clear and sunny day. In the same way, bats have an acoustic fovea within their ears that can analyse small changes in the frequency of sounds within a small frequency range, corresponding to the frequency at which they emit sound pulses.

Knowledge of how the brain computes information using echolocation is helping the blind to see. Scientists have developed dark glasses that are mounted with mini-speakers that emit sound pulses, along with microphones that listen to the echoes and convert the information into a visual landscape. The sensory abilities of our wild friends not only help them to survive, but can assist humans to use sounds to find our way in the dark as well.

Seafarers that swim with sounds

Although they live in water, dolphins are of course mammals and breathe air through their blowhole, which is located at the top of their head. Some types of dolphins must rise to the surface to breathe every 20 to 30 seconds, while others can hold their breath as long as 30 minutes. Dolphins and whales are the only mammals that swim under water and echolocate. But their sound pulses are not like those of most bats. They have found a different way to use sounds to obtain information about their environment.

Instead of emitting constant frequency sounds or pure tones and/or frequency modulated sweeps, whales and dolphins emit clicks, which are brief, broadband sounds of high frequency. We can do it too by sticking our tongue up to the palate and then rapidly disengaging it by opening our mouths. But our ears and brains aren't especially designed to extract the information from the echoes of clicks reflected from an object. Dolphins have no problem doing so. In fact, the work of several scientists has shown that they are extremely good at this.

The average hearing range for humans is about 0.02 to 20 kilohertz (kHz). Bottlenose dolphins hear tones with a frequency up to 160 kHz, with the greatest sensitivity ranging from 40 to 100 kHz. As sound travels four and a half times faster in water than in air, the dolphin's brain must be extremely well adapted in order to make a rapid analysis of the complicated information provided by the echoes. Although the ability to echolocate has been proved experimentally for only a few odontocete species (dolphins and whales that have teeth), the anatomical evidence suggests that all dolphins have this exotic ability.

We vocalise (make sounds) by exhaling or forcing air out of our lungs and through our larynx. Vocal cords in the larynx vibrate by rapidly opening and closing at different frequencies as air flows across them, producing sounds. Our throat, tongue, mouth and lips shape these sounds into speech. A dolphin doesn't have vocal cords in its larynx. Instead, forced air movements over some type of 'vocal folds' within the nasal passage probably produce the sounds they use for either echolocation or communication. Technological advances in bioacoustic research have enabled scientists to better explore the nasal region. Studies suggest that the 'vocal fold' is actually a tissue complex in the nasal region that is probably the most likely site of all sound production. This complex, called the dorsal bursa, includes 'phonic lips' – structures that project into the nasal passage. As air pushes through the nasal passage and past the phonic lips, the surrounding tissue vibrates, producing sound.

As soon as an echo is received after the sound bounces off an object such as a fish, the dolphin generates another click. Like bats, a dolphin can figure out exactly where the fish is and where it's going. The time lapse between click and echo enables the dolphin to evaluate the distance between it and the object; the varying strength of the signal as it's received on the two sides of the dolphin's head enables it to evaluate direction of the target. If the dolphin keeps producing clicks and receiving the echoes, it can extract information about the speed and direction at which the fish is moving. If the dolphin is far from the target it will produce clicks at a slow rate. The closer the dolphin gets to a target, the faster the clicks bounce back and the faster the dolphin sends out more clicks to detect the object that the clicks are bouncing off. By continuously emitting clicks and receiving echoes in this way, the dolphin can track objects and home in on them.

The echolocation system of the dolphin is extremely sensitive and complex. Using only its acoustic senses, a bottlenose dolphin can discriminate between practically identical objects which differ by 10 per cent or less in volume or surface area. It can do this in a noisy environment, can whistle and echolocate at the same time, and can echolocate on near and distant targets simultaneously – feats which leave human sonar experts gasping. Dolphins are also able to discriminate between two metal discs placed in front of them, in which the only difference is the texture on the surface of the backside, the side opposite to the surface that the dolphin is scanning with its echolocation pulse. This ability to detect internal structure is simply mind-boggling. It's not like using sonar as a substitute for vision. It's more like Superman's X-ray vision under water. In fact, an echolocating dolphin can detect a 2.5 cm object from 72 metres away! It's highly likely that the brains of different species of dolphins are tuned to the frequency of sound reflected from the swim bladder of the type of fish they feed on, since this structure is known to produce the main echo – and they can even tell the difference between different species of fish. This might even explain the shape of the head of bottlenose and other dolphins that is designed to focus the energy in the outgoing click and returning echo.

A number of countries have supported a lot of behavioural research on dolphins because of long-standing interest in the marvellous abilities of dolphins and in training them for rescue and for submarine operations. The US Navy, the world's largest sea force, for example, studies the military use of marine mammals, principally bottlenose dolphins and California sea lions, and trains animals to perform tasks such as ship and harbour protection, mine detection and clearance, and equipment recovery.

Whales, with their even larger brains, are cleverer still. According to a recent article published in the British journal *Biology Letters*, some whale species have sonar systems that send out two pings at once, allowing them to detect underwater objects with greater accuracy than even the most sophisticated human technologies. The 'two distinct pulses' of sound are so close together that they are indistinguishable to the human ear, but can be clearly resolved using computer analyses. Using an endoscope lowered through the whale's blowhole into the upper nasal region, scientists were also able to confirm that the 'phonic lips' can produce clicks either simultaneously or independently. Focusing on beluga whales, Marc Lammers of the University of Hawaii and Manuel Castellote of the Oceanografic Aquarium in Spain speculate that the double sonar system holds several advantages for the species. One is greater carrying power – the combined energy of two pulses significantly expands the range over which the whale can detect objects. The fact that each set of phonic lips produces different frequencies also results in a broader spectrum of sound, providing 'advantages for target detection and classification'. It's also possible that 'the beluga may use slight time delays in the production of each pulse to actively control the width and orientation of its echolocation beam'. This exceptional versatility associated with the ability to produce double pulses is shared by other whales in the same sub-order of odontocete cetaceans, such as killer, pilot and sperm whales.

★

Thus we see that echolocation, first discovered in bats, is in fact a widespread strategy adopted by several species of animals to use the sense of hearing in new and interesting ways. Echolocation has surfaced many times over evolution in different species and has taken many different forms. Besides bats, dolphins and whales, some birds, such as swifts and the oilbird, are also known to echolocate. Some modern-day shrews, having common ancestors with bats, can echolocate under water and possibly in air as well. Although they have not yet been well studied, their echolocation signals range from brief clicks to various combinations of complex chirp-like cries and whistles. As with much in nature, there are many ways of solving a common problem – in this case, of using sounds to visualise one's surroundings.

4

Tasting and Touching

[D]ispirited after a dreary winter day, with the prospect of a dreary morrow, I raised to my lips a spoonful of the tea in which I had soaked a morsel of the cake. No sooner had the warm liquid mixed with the crumbs touched my palate, than a shudder ran through me, and I stopped, intent upon the extraordinary thing that was happening to me. An exquisite pleasure had invaded my senses … with no sense of its origin. And at once the vicissitudes of life had become indifferent to me, its disasters innocuous, its brevity illusory.

Marcel Proust, *Swann's Way* (1913)

Dinner is served!

The pleasure of taste starts with the taste buds and ends with electrical signals reaching the reward centres in the brain. Our taste buds are continuously being regenerated every three to four weeks and the brain cells to which they connect are themselves constantly at various levels of activation depending on our physiological and mental states. It's because of these fluctuations that some foods can taste so good sometimes, and really not so good at others.

Animals taste and enjoy their food as much as we do. Like many five-year-olds, they too are finicky about what they eat and don't eat. In fact, the taste system is one of the most important sensory systems in the animal world because it may decide whether an animal lives or dies, for example by successfully warning of, or failing to detect, poisonous substances before they're ingested. It's the guardian of the body and decides what is going to provide the nourishment required to grow and reproduce.

Watch a squirrel closely next time you come across one squatting on the lawn holding an acorn with its two hands and nibbling the insides. You'll see it nibbling away with its teeth quite rapidly. What you don't see is the tongue inside the mouth that is busy manipulating the little bits of food and tasting the ingredients, swallowing what is delicious and even just acceptable (yes, acceptable too can be satisfying!) and spitting out anything that might be unsavoury. Of course, if there's no time to taste and swallow its food, it will simply fill up its mouth until its cheeks pop out, then scramble away at the slightest sign of danger to taste and swallow the food at leisure in a safe location.

All animals have taste buds, including those that live under the water. Every creature in the ocean and in freshwater bodies, as well as on land and in the air, has its very own set of chemosensory structures and special adaptations to detect the chemicals in their food. Cells within their brains are also tuned to detect and process the chemosensory signals. That is to say, every species lives in its own chemosensory world, although it may have substantial overlap with those of other species.

★

This chapter will take us first of all on an underwater journey into a world where the strangest of creatures have survived for millions of years, carefully tasting and selecting their food to nourish their bodies. In fact, there are more varieties of bony fishes (Osteichthyes) than any other taxonomic class in the whole world. They represent nature's many experiments for the design of a tongue in tetrapods (land vertebrates) and their sense of taste and smell. New fish species are still being found and perhaps many await discovery by scientists in the future. Each discovery of a new species opens our minds to new possibilities of sensing our world, of learning about behavioural strategies for adaptation and survival and strange sensory adaptations that defy our wildest imagination.

The taste of catfish

In Montana in the north of the USA, fishponds are frozen solid in the winter. In Louisiana in the deep south, they are icy cold with a few flakes of ice floating on the top. In late December of 1984, Jag took a plunge into these freezing waters in Baton Rouge, Louisiana. He was then a graduate student at Louisiana State University (LSU) and his lab mates and he had placed traps in the ponds to capture foot-long, juvenile catfish for their research on the taste system.* The annual winter meeting of the Association for

* Not on the different ways to prepare a catfish meal that may taste good to us, but on how well catfish can taste *their* food.

Chemoreception Sciences was fast approaching and he desperately needed data to make a good presentation.

He could feel his body getting numb as he groped under the water and was able to scoop a couple of channel catfish, *Ictalurus punctatus*, into a bucket. He looked at his pale and wrinkled fingers with despair and at the smooth and silky skin of two healthy catfish with absolute joy. Once he had trudged across the university farmland to reach his car parked half a mile away, the bodily sensations started coming back and the excruciating pain was something never experienced before. By the time he was in the car and had turned the heat on, he was shivering vigorously and hyperventilating, and he nearly passed out. Little did he realise at the time that lowering the core body temperature by even a few degrees can be life-threatening. All he cared about was getting some answers to the questions he had posed on how a catfish might taste its food.

Why go to all this trouble to understand the mechanisms of taste sensation in a catfish? Around the turn of the 18th century, 'taste buds' was the name given to tiny bumps seen under the microscope by a famous American scientist named C.J. Herrick. These taste buds, typically present inside the oral cavity, were in fact present *on* catfish. Herrick discovered that virtually the entire body of a catfish is covered with taste buds! No wonder that in Jag's lab they used to call it the 'swimming tongue'. Herrick spent a good part of his career proving that catfish indeed use these structures to taste their food. It took another 20 to 30 years before taste bud-like structures were also discovered to be present on the human tongue. Actually our taste buds exist within tiny protrusions or papillae sticking up from the surface of the tongue. You can see these papillae if you stick out your tongue and look closely in a mirror.

C.L. Herrick established the *Journal of Comparative Neurology* towards the end of the 19th century. As the chief editor of this journal, his brother C.J. Herrick published a series of articles on his experiments with bullhead catfish, *Ictalurus nebulosus*. He proposed that catfish are able to detect chemicals leaching out from the skin of prey. Some of his experiments focused on dipping a piece of cotton in water in which crayfish were present and gently approaching the flank of a catfish with it. Even before it had made contact with the skin on the flank, the catfish would immediately swing its head to the side being stimulated and bite the cotton. Herrick concluded that some chemicals oozing out of the cotton were activating chemoreceptors on the flank. His anatomical studies showed the presence of a special nerve found only in catfish that branched from the facial nerve (one of the twelve cranial nerves present in all vertebrates), made a sharp turn and instead of innervating the face, was diverted to the taste buds in the skin on the flank region. This so-called recurrent branch of the facial nerve carried the taste signals to the large facial lobe in the brainstem that is specialised for processing taste information. Nearly 100 years later, Jag, examining a slice of a preserved facial lobe of the channel catfish, discovered here the presence of giant neurons that were over twenty times larger than the surrounding cells. Their large network of dendrites branched to receive signals travelling through the recurrent and other branches of the facial nerve at up to 23 feet (7 metres) per second. This sparse population of relatively few neurons appeared to be well suited to integrate and compare taste information coming from different regions of the body at any instant and 'decide' in which direction the fish should turn to snap at the prey. Their job was to feed information to command neurons that could contract the flank muscles and get the final job of flexing the body, turning the head back and snapping done in a rapid and efficient manner.

For catfish, indeed all fishes, these potent tastants – molecules that stimulate the taste buds – are not salts and sugar, but amino acids that are the building blocks of proteins. This is highly convenient. It so happens that free amino acids are released from the skin of all animals and especially from the skin of an animal that is injured. Catfish are omnivorous, and some species like the channel catfish are predatory and active swimmers. In addition to their taste-sensitive flanks, all catfish have four pairs of long feelers or whiskers, also called 'barbels', which are literally taste organs loaded with millions of taste buds. They can easily whisk them in the water as they 'feel' the flow of amino acids and try to determine their source even if their surroundings are muddy and obscured. In fact the eyes of bullhead catfish are naturally tiny and they can't see much at all. They frequently reside in muddy waters, unlike the channel catfish which prefer clearer waters and use all of their senses, even electroreceptors, to detect and attack their prey and swallow them whole.

In the second half of the 20th century, scientists performed rigorous experiments to understand the role of taste organs such as barbels on the feeding behaviour of catfish. They discovered that by comparing tiny differences in concentration of amino acids in the water around them, catfish can locate prey several hundred feet away in any direction.

Their taste buds are so sensitive that a catfish swimming at one end of a very large pond can detect a spoonful of a solution of amino acids poured into the water at the other end, and an injured animal can be detected from a mile or more away. Inputs from the taste buds on the head and barbels reach reflex centres in the brainstem that coordinate movement. They are wired to allow catfish to make complex computations to calculate where in their three-dimensional space the source of the food is, and how to get there. Research has shown that if the barbels on one side of the body are surgically removed, these catfish become 'blind' on that side and always turn to the opposite side in a circular tank as they swim to find the source of taste molecules.

As well as taste buds outside their bodies, catfish also have them in the more familiar location inside their mouths – indeed all fish need to have oral taste buds, no matter where else on their bodies they may also carry them. A few scientists, including the co-author of this book, have wondered how the two groups of taste buds differ in their sensitivity and specificity to tastants.

By comparing electrical activity generated in the nerves innervating the oral versus the extra-oral taste systems, one is able to compare the similarities and differences between the two systems. Once again, nature seems to have found a very pragmatic solution. The taste buds located on the outside of the body and lip region of the catfish are not responsive to bitter substances, whereas those at the back of the oral cavity are highly sensitive to bitterness – many substances that taste bitter are known to be toxic if ingested. Even in land animals, humans included, the back of the tongue has taste buds that are highly sensitive to bitter substances and are designed to inhibit the swallowing reflex to protect our bodies from toxins. This is why it can be so hard to swallow a bitter pill placed at the back of the tongue. In fact, bitter substances can also go a step further and trigger a regurgitation or gagging reflex.

Some amino acids play a more important role in motivating an animal to feed, whereas others may encourage an animal to ingest more or less of a particular type of food. In catfish, taste receptor cells and the nerves innervating taste buds on the extra-oral (outer body) surface are highly sensitive specifically to some amino acids such as L-alanine that taste sweet to us. In contrast, the nerves innervating taste buds present on the intra-oral surface are highly sensitive to L-arginine and other substances such as quinine that taste bitter to us.

Food on your palate

Thousands of species of fish have scales on their bodies to protect their skin from injury and so do not have taste buds on the outside body surface, their extra-oral taste buds being concentrated solely on the lips. One such species of fish that we are all very familiar with is the goldfish. If you take a close look at goldfish in an aquarium, you will see that they frequently suck little particles from the gravel and then spit out some of them. What they are doing is sampling the water for food particles with a relatively huge organ inside their mouths. This organ is called the palatal organ because it is located on the upper side of the palate. Scientists such as Thomas Finger of the University of Colorado have done extensive studies on this organ and have discovered that it's full of taste buds. Without external taste buds, goldfish cannot detect their food from afar like catfish do. Instead they need to take a mouthful of the water containing the food particles and then sort them carefully using this muscular palatal organ. Once this structure has grabbed all the tasty little morsels, the fish tighten their opercular muscles to reverse the flow of water and spit out the rest of the inedible stuff under high pressure. These movements are coordinated by a relatively large lobe in their tiny little brains. This so-called vagal lobe gets its taste input through a branch of the vagal nerve and is efficiently organised into layers and columns just like the cerebral cortex in humans.

The brains of fish are amazingly adaptable in an evolutionary context, providing us with the results of literally thousands of experiments performed by nature as each species has tried to adapt to its special ecological niche and feeding habits. Some brain structures contract or expand in various ways to make the most efficient use of the incoming taste information.* There are other equally innovative ways to analyse the water around you, some of which don't even involve the taste system. A different chemosensory system will do, as long as the sensory structures are specialised to detect the relevant molecules and the nervous system is wired appropriately to extract the chemosensory information and coordinate the feeding behaviour.

* In some fish, the vagal lobe is shaped like a bizarre corkscrew.

Finger-licking good

The sea robin, one such inventive species, is a bottom-dwelling fish, living at depths of up to 660 feet (200 metres). This fish gets its name from the large pectoral fins which, when swimming, open and close like a bird's wings in flight. Another distinctive feature is the presence of a 'drumming muscle' that makes sounds by beating against the swim bladder. This is likely the sea robin's way of communicating with its friends, much in the same way that humans have used drums for communication over long distances.

What makes the sea robin of great scientific curiosity are the first three rays of the pectoral fins. These are membrane-free and are used for chemoreception. They shine in the ocean with a vast array of vivid colours. A look at the spinal cord of these fish reveals three large swellings – one for each of the specialised rays. The rays function as the fingers and tongue of the fish, both probing and tasting small animals among weeds before putting them in its mouth. The swellings in the spinal cord are tiny little brains just like the vagal lobe in the brainstem of the goldfish that receives inputs from the palatal organ. These structures quickly process a large amount of incoming taste-like information and tell the fish how delicious a particular item in its meal is, and what to do with it.

This ability of the spinal cord to taste food is unique in the animal kingdom, and has been the subject of several studies by curious scientists who wanted to understand the connections of the sense of taste between different species. What they discovered was that not only fish, but spiders and many insects can taste their food by the structures that are most likely to first come in contact with the food. And this in many invertebrate species turns out to be the feet. Flies walking around in circles on a dinner table and fruit-flies, *Drosophila*, doing the same on a ripe peach are literally tasting the food with their feet. This is accomplished by tiny hair-like structures that have a single pore at their ends. Each neuron located at the entrance of the pore is sensitive to a different group of chemicals. Most insects have a series of neurons that respond to salt, sugar, water or amino acids, and special neurons that respond only to key components of species-specific food items. Similarly, spider taste hairs are located on their feet and at the palps (the two appendages at the front of the head), and vibrations from the frantic dragging at the strands of the web by an insect trying hard to escape may be the perfect signal for a spider to dig its feet into a tasty meal. On the other hand, if you apply lemon cleaner to the wall of your house, the spiders will really hate the taste and pretty soon will move well away. To attract mates via the reward of taste, female wolf spiders lay a trail of silk impregnated with their pheromones, which the males taste as they follow the thread.

Next time you are enjoying a delicious meal, think about some of these creatures that are all trying to do the same in their own ecological niche. And just like us humans, they are making full use of what nature has provided them – an abundance of taste buds, an exquisite sense of taste and, as we will see, a sharp sense of smell to motivate them to find the best food in their environment, as long as it exists.

What about smelling?

Not only do catfish taste amino acids, but they smell them as well; and in the olfactory, as in the taste system, there are differences in the types of amino acids to which the receptors in different species respond best. This raises an intriguing question – how does a fish smell amino acids? Amino acids are not volatile substances, they do not easily evaporate in air at room temperature like alcohol, which we can smell from a distance as we approach someone who has been drinking a lot. But then fish don't need anything to evaporate to smell it because they live in water. They just need it to diffuse through the water. In that case, the classic distinction that holds between smell and taste in land animals doesn't really hold for aquatic creatures. Then why do they have smell and taste as two separate senses for detecting chemicals in the water? To resolve this dilemma, we have to consider the behaviour that each chemosensory system controls.

Research by several scientists has shown that the olfactory inputs project to the motivational systems in the brain. Unlike the distant taste receptors on the extra-oral surface of a catfish, olfaction, including ours, may either suppress or enhance the motivation to feed. In fish, especially catfish, the olfactory system is sensitive to a different set of amino acids and other naturally occurring chemicals, such as bile salts, that may partially overlap with the tastants.

Remember when you were a kid and your mum made biscuits or baked bread? The aroma that filled the house was enough to make you want to eat, and you would use your nose to roughly find where the smell was coming from. Of course, knowing your way around the house and to the kitchen, you needed only to use your memory tied to that smell to know where to head. The next action was to grab the warm biscuits and stuff them in your mouth. There the taste system would take over to tell you if each bite was really as good as it was expected to be. You could immediately detect any over-baked or burnt crumbs and might have chosen to spit them out. Of course, some of you probably liked the crunchy crumbs as long as they weren't really burnt and too bitter. So, the olfactory system allows us to detect the food from a distance and motivate us to get to it, whereas the taste system allows us to decide what to ingest and what to discard. In this functional sense, olfactory and taste systems are distinct, that is to say, each one has a dedicated role to play; and these roles are similar across a diverse array of species whether they live on land or in water.

In catfish too, smell, even though it functions in an aquatic environment, is designed and wired within the brain to function as a motivational system. The sensory signals motivate the animal to look for food. Then the exquisitely designed taste system with its four pairs of 'taste antennae' or barbels takes over. The distribution of tastants in the three-dimensional space around the fish is quickly computed within the brain by comparing concentration differences of amino acid molecules, and the fish starts to navigate its way towards the food source. It may go around in a few spirals, a bit like a recently released homing pigeon getting its bearings from a city's landmarks, or perhaps from the direction of the earth's magnetic fields, but eventually it gets to the source. Once there, the oral taste system in conjunction with an acute sense of touch sorts the soft and tasty from the inedible portions as the fish takes big bites from the prey or carcass. The catfish has used all divisions of its gustatory and olfactory systems to find another meal and live for another day.

So we see that smell in all species induces a motivational switch to initiate feeding, while taste coordinates reflexive behaviours associated with ingestion. Of course, we all experience taste-specific hungers – there's always room for dessert, or for ice-cream when strolling in the park – that can influence the motivational on-switch, blurring the behavioural classification of taste and smell at the perceptual level. In the case of taste-specific satiety, specific taste signals can also trigger an off-switch to terminate feeding in a timely manner. When this switch fails to operate at the appropriate time, the result may be obesity, a condition rarely observed in animals with the exception of some overly pampered pets.

Polar bears

Like catfish, other species make full use of all of their chemosensory endowments. Reptiles, however, are not considered to be big on taste, and birds that fly rely heavily on audition and vision and have only a few taste buds on their tongues. Most are thought also not to have a good sense of smell. Vultures, however, are an exception and can smell a carcass from miles up in the air.

Among mammals, polar bears have an incredible sense of smell. Polar bears live deep in the white Arctic land mass, but can swim in and under water. Their body temperature, which is normally 37°C (98.6°F) like ours, is maintained through a thick layer of fur, a tough hide, and an insulating layer of blubber.

Polar bears are marine predators of the Arctic sea ice. The colour of their furry coats makes them virtually invisible to their prey. It's the sensitivity of their sense of smell that together with their camouflage allows them to hunt successfully in the barren environment of the snow dunes and Arctic ice fields.

The nose of polar bears has a rich supply of blood. This may serve two functions – to lose excess heat (even under Arctic conditions, they move slowly and rest often to avoid overheating) as well as to fully oxygenate the olfactory structures deep inside the nasal cavity. They can smell out a whale carcass twenty miles (36 km) away or seals up to six feet beneath the snow and ice. Sniffing first with its nose pointing up to the sky and then punched into the snow, the polar bear closes in on its prey – say, a newborn pup of an Arctic sea lion. Once the sea lion's den is discovered, the polar bear will use its powerful forelegs to break through the snow and ice in order to reach its defenceless prey. On other occasions, bears will find a seal air hole and sneak up on it slowly and sit there until a seal comes up to breathe, at which point they scoop it right out of the hole. During a season of minimal ice extent, a polar bear will also hunt for terrestrial prey such as barnacle geese and nestlings of glaucous gulls.

The habitat of polar bears is the most threatened by global warming and the fracturing and melting of the ice and icebergs that sustain life in a delicate balance under extreme conditions at the earth's polar caps. As climate change continues to dramatically disrupt the Arctic, polar bears and their cubs are being forced to swim longer distances to find food and habitat – and if the cubs don't make it, nor will the species. In the absence of predation by polar bears, the sea lions may multiply and put undue pressure on their own food resources, which will ultimately cause their populations to dwindle as well. In the delicate balance of nature, elimination of any one piece of the ecological network can make a whole ecosystem maladaptive and even non-functional. Thus, adaptations, sensory and morphological, are meaningful not just from the perspective of a single species as observed by Darwin, but from the perspective of an ecosystem as a whole.

The secret touch

The sense of touch is due to the very sensitive neurons that respond to any deformation of the plasma membrane or the skin of a cell. This contact sense is not as well studied as some of the other senses. The sense of touch is such an integral part of our bodies that we always take it for granted. We don't associate it with anything special because there are no obvious sensory structures or organs associated with it, such as the tongue for tasting and the nose for smelling. In fact, there are several different types of tiny sensory structures under the skin. Many animal species have an exquisite sense of touch and they sure know how to use it.

Whiskers are an integral part of many animals' sensory repertoire. Rats and all other rodents have them, cats have them, dogs have them, buffaloes have them, sea lions have them and even bats have them. Some monkeys have them, but most primate species, including humans, appear to have lost these interesting structures. Perhaps in humans the function of whiskers is taken over by the ridges of skin on our fingers, which are highly sensitive to touch. In humans, this sense is referred to as the haptic sense. To illustrate just how sensitive the fingertips are in relation to other areas of skin, close your eyes and briefly rub your fingers on different grades of sandpaper. You can relatively quickly order the mixed up pieces of sandpaper according to their roughness.

However, since most other animals need all of their limbs to walk, they are in greater need of alternative touchy-feely structures. Research on the brains of animals that employ whiskers for this function has shown that a part of their sensory cortex is devoted entirely to the sense of feeling they provide. This part of the cortex has come to be referred to as the barrel cortex because each whisker is represented in a barrel-shaped field or grouping of cells that receives information from it. How the brain extracts and interprets sensory information from whiskers is not yet well understood, but engineers are already designing robots with whiskers as well, since they are highly adapted to helping them find their way around a room without crashing into walls. Since whiskers are flexible, they are like touch at a distance and keep the head and body from impacting against a surface. They are great for damage control when a robot or an animal is exploring a new environment, especially in the dark.

Reach out and touch someone

Cockroaches have whiskers too, but these are found at the back instead of the front. Their whiskers are on special structures called cerci. These are like distant touch receptors, since they are highly sensitive to air flow. Experiments and calculations have shown that the cerci of roaches and crickets are some of the most sensitive sensory structures in the entire animal kingdom and can detect movement of air particles of less than a micron.

Among underwater mammals, manatees get the prize for developing some of the most sensitive touch sensors. Recent research suggests that the manatee's tactile sense is so finely tuned that the animals may experience 'touch at a distance' – an ability to 'feel' objects and events in the water from relatively far away. In recent studies, marine biologists Roger Reep and Diana Sarko at the University of Florida in Gainesville found that the giant mammals are covered with special whisker-like hairs that act as sensors. These sensors are known as tactile hairs and they form a kind of sensory array, possibly allowing manatees to detect changes in water current, temperature, and even tidal forces (and surely tsunamis) from thousands of miles away. When a hurricane is coming, they get out of the area fast.

Manatees don't have very good eyesight, but they can still find their way in a maze-like network of waterways, such as those found in an area called Ten Thousand Islands near Naples, Florida. Scientists think that one of the possibilities is that they're able to use their covering of tactile hairs to detect the movement of water and tell them where they are in the environment, and therefore they're using their sensory hairs as a navigational tool. Manatees certainly have more brain space dedicated to the sense of touch than many other mammals do. The brain regions associated with touch are 'especially large' in manatees – as large as or larger than in animals known to be particularly sensitive feelers, like star-nosed moles.

The often overlooked haptic sense is far more intricate and important than we might like to think. Even in humans, it's what helps us to navigate around our environment safely, and it's touch that delivers such an exceptional advantage through our sensitive fingertips. In fact, touch is so central to almost every animal, from cockroaches to humans, that studies have shown that hearing deficits in adult animals result in a conversion of their brain's sound processing centres to respond to touch. And as we shall see in the case of bats below, even when animals might not be thinking about touch, it's a sensory system that controls unconscious reflexes vital to everyday life.

Bat acrobatics

Have you ever tried stroking the sole of your foot? In addition to the giggling it might produce, stroking the sole of the feet also makes the toes flex inward. This is used clinically to test if the motor reflexes in the spinal cord are working after someone has been in an accident. Damage to the motor pathways produces an extension of the toes rather than the normally observed flexing in response to stroking. This is known as the Babinski sign. The normal reflex helps us grab the surface we are walking on. For humans, therefore, touch plays an important role in simply being able to walk without falling.

Bats have a heightened sense of touch on many parts of their bodies, from the whiskers to the delicate skin on their webbed wings to the under-surface of their feet. The ability of bats to flex their toes to provide the grip needed to hang upside down in trees or from cave roofs is most likely a simple reflex from stimulation of the under-surface of the feet. This touch-based response is so ingrained in bats that bat pups are even born feet first and, perhaps unsurprisingly, they have exceptionally large feet. New-borns of most species, including humans, have relatively large heads; bat pups, when they are born, have both a large head and large feet to match, which they use to hang on to their mothers as soon as they are born. Not being able to do so would mean a fall and certain death among the creatures inhabiting the bottom of the cave. And once they are fully grown, they must use the flexing reflex each time they have to land on a tree limb or a protrusion on the wall of a cave and hang there for hours.

In March 2009, Henry Fountain of the *New York Times* reported: 'Sticking the landing may be hard for gymnasts, but at least they finish right-side up and have gravity on their side. Consider what bats have to do: because they roost by hanging from their heels, they must land upside down against a cave ceiling or foliage. And on their approach they fly upward, against gravity.' Daniel K. Riskin of Brown University and colleagues have now shown how bats do it. They perform a flip, and in some cases a twist, and make a two- or four-point landing. Bat hind limbs are thin and light and very fragile, and landings must be executed safely. The severity of the landing may be a function of roosting location. Using a high-speed video camera and a contact plate to measure the forces, Riskin studied the landings of three bat species, two that roost in caves and one that roosts in trees. The average peak force was nearly four times the body weight in tree-roosters, but landing on all fours reduced the stress on any one limb. Also, since foliage has some give to it, a hard landing won't hurt and might even give the bat time to make sure it has a good grip. The cave-roosters, on the other hand, landed more softly, yawing to one side during the flip to end up on their hind limbs only. On hard cave ceilings, they must land softly to avoid injury.

Humans are bestowed with some supreme senses and abilities that we consciously and unconsciously use to navigate and survive in our outer worlds. We use them to literally make sense of the world we live in. As we have seen in this first part of the book, animals do the same, except that many of them are endowed with some unique capabilities and stimulus detectors that go along with such capabilities. These include the ability to use electric and magnetic fields and the softest of sounds as alternative strategies to visualise the environment and exploit the stable and familiar cues to navigate, survive and thrive. Other species that share our senses use them in ways that may appear strange to us. What research has shown is that nature finds a solution, sometimes even the same or a similar solution multiple times over an evolutionary timescale, to achieve its goal of allowing an animal to adapt to its ecological niche.

The brain and its neural circuitry then capitalises on the different sensory adaptations and processes the relevant information in the most efficient manner to generate and coordinate goal-directed behaviours.

Part II

Surviving

5

Alarming Behaviour and Survival Strategies

And we are here as on a darkling plain
Swept with confused alarms of struggle and flight …
Matthew Arnold, 'Dover Beach' (1867)

Years ago in Vienna, Karen and some college friends observed something they'll never forget. As they sat quietly on a bench taking in the beauty of the Habsburg Palace grounds and the pond in front of them, the peace was suddenly interrupted by a plop, then a frantic splashing and fluttering of wings. Next, a duck that had been across the pond swam to shore in front of them and carefully set down a baby bird that had fallen from a nest in a tree limb protruding out over the water. The duck waited to make sure the nearly-drowned fledgling was alright and the mother had come to its side. Then the life-saving duck swam back to the other side of the pond.

Whether it's an accident, a possible predator, or a natural disaster like an earthquake or a tsunami, animals have long been known to perceive possible peril and to notify and protect each other. Let's now look at the variety of alarm signals that animals (and sometimes even plants) send out. Let's explore some of their ingenious survival strategies. And finally, let's try to answer the question of what signals the animals of Yala National Park might have sensed and communicated that saved their lives during the horrendous tsunami of 2004.

Alarm! Alarm! Alarm!

If we're canoeing down a river and see a beaver slap its tail on the surface of the water, the odds are that the beaver is warning of danger – and that we are the perceived danger. If a white-tailed deer gives out a quick 'deer whistle' and flashes its tail-patch in front of other deer, it's saying: 'Let's get out of here.' A fussing squirrel is broadcasting its own concern about a possible predator. And no one can mistake the meaning of the raucous shrieks of a crow, the hissing of a cat, the sound of a rattlesnake's tail, or the growling of a dog – fangs bared. Animal alarms can protect a specific species, or they can involve groups of animals that know each other's signals – such as the zebras, gazelles, giraffe and other animals stalked by the lionesses of the African plains. Plants can be part of the alarm system as well.

Scientists have known for quite some time that bacteria give off alarm signals, and they are doing increasingly sophisticated research to show how plants do so as well. Under water, seaweeds send out their own sea-alerts. Studies have shown that single-cell organisms such as some green algae (e.g. *Euglena* and *Chlamydomonas*) give off certain chemicals to warn of danger, and research is ongoing to find out what types of signals – including seismic and electrical – are also a part of a plant's alert systems.

Scientists have discovered that plants do indeed generate electrical signals that can be picked up by other plants, and by insects and perhaps other animals. Since 1873 when British physiologist Sir John Burdon-Sanderson detected bioelectrical activity following stimulation in plants, a great variety of research has been carried out to better understand how environmental stimuli, often referred to as irritants – such as temperature change, humidity, light, and wounding – affect the electrical signals within and the electric fields that emanate from a plant. In one study, researchers hooked electrodes to the leaves and stem of a philodendron plant. They had ten human subjects come into the room one by one and stand by and/or touch the plant. When one subject tore a few leaves off the plant, the electrical activity in the plant jumped way up. The next day all ten subjects went one by one back into the room and stood next to the plant. When the person who had damaged the leaves of the plant the day before came into the room, the plant's electrical activity again zoomed to a high level. The researchers then put other plants into the room with the philodendron. The next day, each of the ten subjects went into the room with the group of plants. When the 'leaf-tearer' walked in, the philodendron sent out what appeared to be strong warning signals. The other plants must have picked up this alarm, because when the subjects visited on the fourth day, all the plants in the room increased their electrical activity – almost in unison – when the 'leaf–tearer' entered the plant area.

Many, but not all, of these bioelectrical signals resemble processes typically associated with nerve potentials. These potentials are generated by the opening and closing of positive (sodium, potassium and calcium) ion channels in a specific sequence. When the potential reaches a threshold value, thousands of positive (usually sodium) ion channels open and close synchronously, leading to a rapid flow of charge in and then out (via potassium ions) of the cell. This makes a cell's membrane potential fluctuate, and the fluctuation is referred to as an 'action potential'. Electric potential changes have been studied in great detail in the giant cells of *Chara* and *Nitella* plants. In these and other plants, waves of excitation travel from the top to the stem to the root and back to the top of the stem. But they don't travel at identical rates because the speed of electrical transmission is dependent on many internal and environmental factors; the latter include the irritants we mentioned as well as specific electromagnetic and gravitational fields.

Scientists are going even further to prove that there is another level or scale of electrical activity in addition to an action potential; they have shown an electrical signal that they call a 'system potential'. Increasingly advanced and detailed electrophysiological research was done in 2009 by scientists at the Max Planck Institute and the Justus Liebig University of Giessen in Germany. Tiny filamentary electrodes were inserted through stomata openings – minute openings in the leaf surface through which plants regulate evaporation and the exchanges of gases – directly into the inner leaf tissue, and then placed onto the cell walls.

With such delicate electrodes, scientists were able to discern an electrical signal they called 'system potential' that is induced and even modulated by wounding. If a plant leaf is harmed, other plant leaves are notified of this danger. According to the report: 'If a plant leaf is wounded, the signal strength can be different and can be measured over long distances in unwounded leaves, depending on the kind and concentration of added cations (e.g. calcium, potassium, or magnesium). It is not the passive transport of ions across cell membranes that causes the observed changes in voltage transmitted from leaf to shoot and then to the next leaf, but the activation of so-called proton pumps. "This is the reason why the system potential we measured cannot at all be compared to the classic action potential as present in nerves of animals and also in plants," says Hubert Felle from Giessen University.' The system potential is more sensitive than the action potential and can carry different types of information at the same time. The scientists found this system potential operating in all the plant species they studied, including tobacco, maize, barley, and the field bean (*Vicia faba*).

Scientific study of the numerous ways in which plants communicate is just beginning. Some animals, especially insects and spiders, pick up on the plant alarms, and of course the insects, spiders and other arthropods* have many alarm mechanisms of their own. Although these may include acoustic, vibrational, visual, electrical, tactile or chemical signals, most of the research has centred on the latter two, with particular attention on pheromones – as

* Arthropods include crustaceans, insects, spiders, and many-legged bugs such as centipedes and millipedes – which don't really have a thousand feet.

given off by insects such as bees or ants, fish and other animals. According to Bert Holldobler and E.O. Wilson, authors of *The Super-Organism*, ants are the insect geniuses of chemical communication:

> Their bodies are crowded with exocrine glands employed to manufacture pheromones, the chemical compounds used as signals … The ants have enhanced the chemical channel in several ways variously by mixing pheromones from multiple glands, by giving separate meanings to different concentrations of the same pheromone and by changing meanings according to context. They have simultaneously added auxiliary signals of touch and vibration.

Ants may be the geniuses, but yellow jacket wasp squads can be among the scariest. If you squash a yellow jacket within 20 feet (about 6 metres) of its nest, within fifteen seconds you will be surrounded by angry troops coming to the yellow jacket's aid. They have picked up on the pheromone (and perhaps other vibrational) alarms of their fallen comrade. This type of information is being used by some scientists who hope that by studying the alarm pheromones that insects use to defend themselves from predators, we may be able to use these compounds to disrupt communication among insect pests, or even other animals.

From bugs to worms. Those preparing for a fishing expedition are known to go into the night air looking for worms. Odds are, the worms are broadcasting the intruders' presence. Earthworms of all varieties secrete a mucus to help them slide across the soil, as well as to navigate and even fix patches in their burrows. When danger arrives, the alert earthworm sends out an even larger amount of mucus with a danger-alert pheromone in it. Other worms that encounter even a drop of this signal quickly move out of range. (Different species of animals might have received the worm warnings too.)

Some worms 'come out of the soil as if they are running', according to neuroscientist Kenneth Catania of Vanderbilt University in a fascinating article in the *New York Times* Science section on worm grunting – a unique way to get worms for fish bait. In worm grunting, or worm fiddling or charming, a wooden stake is driven into the ground and then the top of the stake is rubbed with a flat piece of steel to make a grunting or snoring sound. If conditions are right, hundreds of worms will rush to the surface of the ground. Catania has shown that the frequencies of the vibrations of the studied worm grunting activities in Florida are very similar to the sounds that the predator eastern American moles make as they dig and move through the soil. So when the worms hear the grunting noises they think the moles are after them, and rush by the hundreds out of the worm burrows and into the clutches of human predators waiting to go fishing. While there's no research on the alarm pheromones that these 'running' worms are giving off, it's reasonable to conclude that they, like the other worms, are giving off danger-alert mucus.

While some alarm pheromones are species-specific, most are shared among many species. There are numerous advantages to this – one species would, for example, be able to interpret and defend against the alarms of others. Most of all though, diverse species could all benefit from an alarm announcing a common enemy or a natural disaster like a hurricane, sandstorm, earthquake or tsunami. Many types of animals send out alarm substances – skunks even spray theirs. Ants and some wasps 'tag' their enemies with the alarm chemical so others will know they are dangerous. Very tiny creatures such as aphids and mites send out their own brand of pheromones, as do water-dwellers like sea anemones, tadpoles and schooling fish. Small mammals like rats and shrews send out warning odours. Hyenas and deer are among many large animals that also give off alarm substances.

Very likely humans do as well. Throughout the ages we have talked about the smell of fear coming from a person. Many humans claim to be adept at picking up such an odour. Certainly, dogs that are known to have extremely sensitive olfactory systems pick up all kinds of odours and are thought to react aggressively to people giving off the 'fear alarm'. Pheromone and vibrational studies regarding human emotion are just beginning to emerge, and are sometimes challenged by those who can't imagine that humans would indulge in such 'animal behaviour'. That mindset, of course, is one reason why many *Homo sapiens* aren't attuned to the animal alarm signals being given off by various species around them.

Animals often face a dilemma when danger is near – how to warn friends and get help without giving out information to the predator. A quiet text message to the police without alerting the intruder might be a human strategy. The European robin's strategy is to get attention by emitting pure-toned high-pitch alarms that fade so quickly that the potential predator, such as a hovering hawk, can't locate the source of the alarm. For their alarm calls, several squirrel species, such as the Richardson's ground squirrel in Canada, resort to ultrasound which is inaudible to most ears outside of their squirrel group. Calling out for help on a different 'radio frequency' than most enemies can pick up also works well for a number of insects, arthropods, bats, perhaps some plants, and most likely for many more animals that humans just haven't been able to hear or pick up on our technological instruments.

Another defensive strategy is to give out a call that draws predators nearer – so the actual location of the predators is known and they may even become distracted so that the potential victim can get away. Yet another ploy might be for a small bird, for example, to give out a signal so that an enemy like a hawk would swoop down to a certain area and find a mouse there instead (an equally satisfying meal), as the signalling little bird flies off unscathed. And then, well known to many, there is the mother bird who can make a great deal of noise or pretend to be injured, so that all attention is on her as she moves farther and farther from her nest of baby birds. Observant humans have been saying for thousands of years that animals send out and understand very sophisticated alarm signals. Scientists have started proving just that.

Different yells tell of different enemies

Is it a bird? Is it a leopard? Is it a snake? Those who know how to listen to the warning calls of African vervet monkeys can answer that question. In 1980 R.M. Seyfarth and his colleagues carried out a groundbreaking study that demonstrated that vervet monkeys give out different alarm calls to warn of specific types of predators. The researchers made recordings of three different types of warning calls that these monkeys were sending out and which had been connected to specific predators nearby. Then, when no actual predators were around, the researchers played these recordings. When they played the warning call that monkeys used to warn of leopards, the monkeys within range of the call rushed to get into trees. When the vervets heard the 'eagle alarm', they all looked up into the sky. And when they heard the alarm call usually sent out to warn of snakes, the monkeys immediately looked down and around for snakes.

The tiny black-capped chickadee, a bird species prevalent throughout the eastern United States, is named after the 'chick-adee' sound it makes. A 'dee' is added to the call when danger looms, changing the call to 'chick-adee-dee'. Proximity and degree of danger are communicated through the additions of even more 'dees': more than twenty have been recorded. That's a lot of 'dees' for anyone to count, but the nuthatch knows what that signal means, so does the titmouse and probably many birds and other animals do as well. The nuthatch and titmouse, small birds that often forage with the chickadee, usually respond in specific ways, depending on the nature of the 'dee' alarm-call.

Christopher Templeton and fellow researchers at the University of Washington analysed the chickadee warnings about fifteen different types of predators. Besides their numbered 'dees', the birds sent out a high-pitched 'seet' sound when they saw a predator flying nearby. Alerted birds all around either dived for cover or froze in place so as not to be noticed and to perhaps wait for further information or instructions. When these came, the number of 'dees' sent out denoted the type and distance of the predator, and seemed to signal the best response. With more 'dees', more chickadees and other birds showed up to harass the enemy by dive-bombing it or shrieking in its face to get it to leave the area. Such a mobbing response to alarm calls is common for a number of bird and other species. The research of Templeton and many others is helping humans become more alert to the extraordinary number of meaningful sounds and signals constantly going on around us.

One interesting alerting signal picked up by neuroscientist Lesley Rogers and her colleagues at the University of New England in Australia has to do with the look of the Australian magpie – the look in the magpie's eyes, to be precise. The researchers found that upon noticing a predator, the magpie signals whether it's going to flee – by fixating on the predator with its left eye – or move closer, perhaps for further inspection – by fixating with its right eye. Rogers associated this finding with research on the human brain that shows that the right hemisphere (to which the left-eye information goes) processes novel and potentially threatening information. The left hemisphere is more prone to carry out information analyses – hence the right-eye investigative stare.

Different dialects, different terrain

Even those humans who speak the same language may sometimes struggle to understand each other. As a Peace Corps volunteer a number of years ago in the north of Somalia, a former British colony, Karen (along with the other young American volunteers) was surprised to find that she couldn't make out some of the English dialogue in the British movies that were screened.

Scientists are increasingly showing that it's not just humans who develop different dialects. In their study of the alarm calls of six different colonies of Gunnison's prairie dogs near Flagstaff, Arizona, Bianca Perla and Con Slobodchikoff of Northern Arizona University found that each colony had its own dialect. And colony members sent alarm calls in their specific dialects. It seems that dialects occur because of group isolation, differences in terrain, and variations in weather as well as habitats. For example, sound travels differently in desert areas than in places with a great deal of foliage. Prairie dogs in mountain areas may incorporate a few echoes in their vocal deliveries. And the humidity impacts on the quality as well as the frequency of tones sent out. Eventually, as in the case of humans, a prairie dog's 'accent' identifies it as the member of a particular coterie (group).

Like the vervet monkeys, Gunnison's prairie dogs have an extremely sophisticated alarm system. A bark that warns of a human nearby sends the little prairie dogs scurrying to their burrows. A bark warning of a hawk makes all the animals look up, but only the ones in the path of the hawk rush to safety. If the 'watch-out-for-the-coyote' bark is heard, the small rodents all run to the lips of their burrows and watch the coyote to see what to do next. If a threat is urgent, the barking rate is increased directly in proportion to the speed of the oncoming predator.

In a recent follow-up study, Slobodchikoff and J.K. Frederiksen found that Texas black-tailed prairie dogs showed the same type of discriminating predator information in their calls as the Gunnison's prairie dogs had demonstrated. This time the research regarding humans became more detailed. Similar to the Arizona prairie dogs, the black-tailed prairie dogs gave out different calls depending on the size and shape of the humans advancing towards their territory and gave different signals based on the colour of their clothing.

Not only this, the black-tailed prairie dogs also sent out different calls depending on whether the human approaching the colony was a 'good guy' or a 'bad guy'. The 'good guy' experimenter had fed the prairie colony members every day during a three-week period, whereas the 'bad guy' experimenter had fired a 12-gauge shotgun into the ground nearby once daily for five days. While it seems that prairie dogs are more aware of different aspects of human appearances and motives than we could have imagined, perhaps it shouldn't be surprising that they have alarm calls based on the level of danger a human may pose, since people have been hunting these prairie dogs for hundreds if not thousands of years. These and many other recent scientific experiments show that animals are much more aware of and affected by their surroundings, and are able to communicate much more to each other, than scientists had previously thought.

Camouflage, bluff or outmanoeuvre

Some species hide out, while others stand out – for a reason. Any smart animal has a list of survival concerns. They don't want to be found by an enemy, attacked, captured or eaten. Of course, if you don't want to be seen, find a good hiding place or use camouflage. Nature provides extraordinary examples of ways in which animals are difficult to spot because they blend in so well with their surroundings. And when the surroundings change, as with the change in seasons, an animal's coat or skin colouring often changes accordingly.

Sometimes the change in surroundings is man-made. In a well known and sometimes debated story about the change in colouring of the peppered moth in late-19th-century England, sooty cities like Manchester were the culprits. As mills gave off dirty black smoke and cities became more industrial, the predominately white moth with black speckles stood out clearly against its sooty background. So over a relatively brief time, the moth is said to have changed to a dark grey or black variant that would be more difficult to see in a darker and more dismal environment.

An animal's colouring can also be affected by the type of predator. Devi Stuart-Fox and colleagues at the University of Melbourne in Australia found that when responding to two different predators, the Smith's dwarf chameleon behaves in the same manner, but varies its colour. When a predator is nearby, the chameleon stays perfectly still on a branch or stalk and rapidly changes its colour. But the chameleon's colour matches its surroundings much more precisely if a predator bird is close by than if a threatening snake is around. Why the difference? The researchers surmise that since the snake's ability to distinguish colour is so poor, the chameleon doesn't have to work as hard creating the perfect camouflage shade. Assuming that changing colour has a physiological cost, the chameleon doesn't put out any more energy than necessary to hide itself from the snake.

Yet, every once in a while even the cleverest camouflage can go awry. Some friends of Karen's have an albino squirrel living near their house. Grey squirrels are prevalent throughout North America and this particular individual thinks like a grey squirrel. Consequently, when disturbed, this beautiful white squirrel runs up the grey-brown trunk of a nearby tree. It then stops right in front of the humans and stays completely still – obviously convinced that they can't possibly see it. For years now, this squirrel has proudly hidden behind camouflage it doesn't really have, and one can only marvel at its longevity. The theme running through this book is how amazing, clever, and survival-oriented animals are, but it's nice to know that nature's not perfect either.

Then there are the animals that really do want to be seen. They have brilliant colours like red, orange or yellow and have some kind of harmful or noxious chemical to harm or repulse any predators. The coral snake in Florida is a deadly example. And even deadlier is the South American arrow-poison frog, decked out in brilliant patterns of red, white or yellow on backgrounds of black or electric blue. The indigenous peoples of Colombia used to use this frog's poison on their arrowheads to paralyse and kill enemies. Many animals that flaunt their bright colours have a chemical poison that can harm the attacker. Or they are mimicking other colourful animals that have a harmful chemical of some type: these unsavoury look-alikes are bluffing – acting like they're a threat to potential enemies. Bright and obvious colours announce to potential predators: 'Stay away from me. I'm full of poison, or a terrible taste or odour.'

Other animals have their own way of bluffing by puffing. They expand their body size in some way to look much larger. Some frogs can suddenly seem twice as big just by filling the loose skin around their throat with air. Several varieties of birds puff up their feathers to appear much larger – hoping to discourage those that might attack. Blowfish blow up their spiny bodies with water or air to scare off enemies. They not only look frightening, but their defence system might work in other ways as well. In one documented incident, an unsuspecting moray eel ended up with quite an enduring mouthful when a blowfish it had snapped up expanded in survival-mode. Engorged, it became too large and too painful for the eel to swallow. Predator and prey couldn't disengage – they were stuck with each other, destined to swim around as an unwilling and desperate pair in an uncompromising position.

There are also snakes with puff-up parts like the poisonous adder and the hooded cobra. Then there's the notorious snake that strikes fear into the heart of anyone who has lived in Africa – the spitting cobra. Karen was introduced to an old Somali man who in his youth had been blinded by the potent venom of this snake. The Mozambique spitting cobra combines a threatening posture with deadly poison that can be discharged as far as 9.75 feet (3 metres). To appear large and threatening, the cobra rears its head and the front third of its long body high into the air, flattening its neck to appear even larger as it sways back and forth in the most menacing manner possible. If this doesn't scare off an enemy, the snake strikes at the eyes of a predator and with force and precision spits out a potent nerve poison that can blind its victim, like the unfortunate man Karen met. The snake rarely has to bite because enemies are usually discouraged from getting too close by its threatening stance and deadly reputation. But it can bite, if it must. And there's yet another trick in the cobra bag. If the enemy persists and tries to tear at the spitting cobra, it will often play dead in hopes that the attacker will lose interest and go away.

Nature has many strategies for attack and for defence. To resist attack, there are the fighters such as attacking wolves, lions, tigers and bears. There are those that flee. Some take flight on land, sea or in the air. Some fliers turn into attack mobs, like wasp brigades or the bombarding bird squads that gang up against a predator. Other animals protect themselves with hard shells or sharp quills. There's the gigantic hairy spider that throws out sticky hairs to repulse and entangle a foe. There's the Virginia opossum that plays dead.

And there's the 'hot-tailed' California ground squirrel. Scientists recently were surprised to learn that this squirrel actually heats up its tail to protect itself against a specific predator – the rattlesnake. Squirrels have long been known to use vocalisations to keep unwanted birds and other mammals at a safe distance. Ground squirrels also tail-flag – wave their tails – to warn of danger and even to ward off a predator. They always do tail-flagging around snakes, because snakes don't pick up airborne sounds. However, it's also known that rattlesnakes are sensitive to the infrared end of the light spectrum. Using an infrared camera, innovative researchers Aaron Rundus and Donald Owings at the University of California found that the ground squirrel has a remarkable infrared weapon. When confronted by a rattlesnake, it heats up its tail and waves it in a harassing manner to appear larger and more threatening. But the 'hot-tail' is used as a defence only against that particular infrared-sensing snake. The squirrel doesn't heat its tail around a gopher snake, it only tail-flags; unlike the rattlesnake, the gopher snake doesn't possess an infrared sensor.

Barbara Clucas, another researcher at the University of California, documented the California ground squirrels tricking their rattlesnake nemeses in another way as well. They chew skin that has been shed by a rattler, then lick the saliva with the snake smell all over their bodies and the bodies of their kittens. Clucas and her colleagues found that the snake scent didn't seem to bother other squirrels, nor did it keep away fleas. The main purpose of smelling like a snake seems to be to ward off being eaten by a snake and perhaps also to scare off other predators. Since the ground squirrels spend a lot of time underground, being covered by snake odour may well make their burrows seem a much less attractive proposition to hunters and scavengers, helping the squirrels to sleep more soundly. California ground squirrels aren't the only ones known to coat themselves with scents of other animals as protection. Rats apply weasel odour and hedgehogs wear toad-skin 'perfume' to avoid being caught by their enemies.

There are a variety of species that if captured can still make a break for it. Literally. Lizards that let their tails or limbs break off in the captor's clutches are able to escape and soon grow back those missing body parts. One lizard, the African blue-tailed skink, uses its brightly-coloured tail as a lure to distract predators from its head and the rest of its body. When a predator attacks the twitching bright blue tail, the tail breaks off and continues thrashing around for a while, further distracting the predator while the skink hurries away to safety.

Many species, especially amphibians and certain fish, can regenerate different parts of their bodies. The zebra fish, for example, is known to replace its scales, fins, retina, spinal cord and part of its heart. Starfish can replace arms and even larger areas of the body. Mammals regenerate body parts as well: deer can regrow their antlers at the rapid rate of up to 2 centimetres a day. All mammals can regenerate the liver; in humans, if three-quarters of the liver is removed in surgery, the organ can regain its original mass in two to three weeks. When we cut ourselves, the healing process around the wound is another example of regeneration. In fact, the tip of a human finger to the joint can often be regrown.

Regeneration usually occurs by some of the mature cells at the site of a wound turning back into a clump of immature cells, a blastema, that regrows the missing part by going back to the embryo's genetic growth-plan for this part of the body. Coordination is established between the mature and immature wound-site cells to replace a missing part.

Medical research has begun focusing on the secret of regeneration. Salamanders are especially good subjects to study because they have a very strong regenerative ability. A salamander can replace an amputated limb in less than a month. It can also regenerate its tail, its upper and lower jaws, the lens and retina of its eye, and its intestine. Not only that, neuroscientists have taken out the brain of a salamander, ground it up, and put it back in its cavity, and soon the salamander is able to function quite well again. The brain connections grow back at a startling rate.

Since all species have a regenerative capacity, researchers are studying the genetic and environmental components involved, with much optimism for future therapies. Scientists have different opinions as to whether specific cases of molecular change and cell growth are 'regenerative' or 'additive'. Both seem to happen in the brain – neuronal fibres and connections have been shown to regenerate. At the same time, throughout our lives – if we use our brains vigorously – we can add new cells and neuronal connections in places where they have not been before. We are already using recent knowledge of neurogenesis – the continuous lifetime growth of brain cells and connecting tissues – to aid with overcoming stroke and other brain damage. Another medical research area is that of rapid replacement of healthy blood.

The Texas horned lizard keeps its tail, but can sacrifice up to a third of its blood to defend against an attacker. This lizard's rough-skinned, toad-like appearance, mottled camouflage, sharp body spines, and a spiny 'crown of thorns' on its head render it a fascinating blend of unsightly and majestic. All these body defences discourage most would-be attackers. However, if an attacker persists, the horned lizard will spit out a focused laser-like stream of blood from the corner of one of its eyes. The blood can be squirted as far as 6 or 7 feet (almost 2 metres) and has a terrible-tasting chemical that is sure to repel almost any wolf, dog or coyote.

Cockroaches and coyotes: the ultimate in survival?

It has long been said that if humans are ever stupid enough to destroy our planet, it would be the cockroaches that would survive and start a whole new zoological cycle. Several years ago, when western coyotes were spotted in Rock Creek Park in Washington, DC, Karen was told by a scientist that coyotes are equally adaptable and prone to survival under the most difficult circumstances. The discussion of the western coyote suddenly showing up and thriving on the East Coast prompted an article in the *Washington Post* that mirrored our colleague's praise for coyote savvy. (There was a follow-up letter, however, saying that if the coyote was so smart, how come he still hasn't caught the roadrunner?) Are there any survival strategies humans can learn from cockroaches, coyotes and other animals – the roadrunner, perhaps?

We certainly know that animals that survive over the long haul have great adaptability and are able to change dietary and mating behaviours as well as habitats. Scientists are also surprised to see how rapidly genetic material in animals can change – this has been easiest to observe in their physical characteristics. For instance, the Darwin finch on the Galapagos Islands significantly changed its bill size to offset the threat of intruding competitors for the same food source; its bill grew noticeably larger within a few generations – just two decades of time. Now that particular type of finch can crack and eat much larger seeds than it could previously. Even more amazing, scientists have learned that it may not even take a generation to change DNA. Recent genetic research, such as that on circadian (daily light and dark cycle) rhythms, as well as on the biochemical oxytocin, shows that our DNA is not just an unchanging 'map' that we're born and stuck with. Not only do our thoughts and experiences influence what genes are expressed (taken from dormancy to active duty), but our thoughts and experiences can actually change our genes and our DNA. So are cockroaches and coyotes really as adaptable as their reputations suggest? While there may be equally adaptable animals, these two species have qualities that render them at the top of the survival-proficiency list.

Although cockroaches are repulsive to most of us, the more that researchers learn about them, the more incredible these insects become. For impressive starters, studies show that cockroaches can live up to two weeks after being decapitated – that's two weeks without a head. German cockroaches have been found to survive without food for a month or water for a week. Even roaches that still have their heads can go as long as 40 minutes without breathing, and females can have up to two million babies a year. Cockroaches are relatively invulnerable to radiation and can tolerate about 100,000 times the dosage of radiation that humans can. This is because the cells in a cockroach's shell covering are difficult for radiation to penetrate quickly. Other insects also have similar protection. Eventually, cockroaches will be affected by nuclear fallout – but not nearly as rapidly as humans and other animals.

Cockroaches have been around for more than 300 million years, and are capable of living in a wide variety of environments. Presently there are estimated to be more than 5,000 species found all over the world, including the polar regions. These ubiquitous insects form and increase their numbers in an emergent way – their self-organising groups don't need a directive from the 'central cockroach office'. Instead they extract information from the environment and quickly organise their behaviour to fit the requirements of a particular area. If their familiar surroundings change – for example, there's a multi-year drought – the innovative insects change their foraging and reproducing habits accordingly. They can also move on to other territory, although they rarely have to because they adapt so rapidly.

Cockroaches can move fast if they have to – to a new home or just away from a predator or other threat. Should alternative methods such as hiding or keeping still to escape danger not work, they have an uncanny strategy of zooming away with a precise diagonal move that usually confounds their enemies. Since cockroaches usually forage at night with the help of antennae through which they detect odours, not as many predators are around. Cockroaches get nourishment from an endless array of food sources, some of which repel humans, and they leave information for other roaches through their dropped faeces, pheromones or other chemical messages, as well as responding to and giving out many types of airborne and electromagnetic vibrations.

The spiny legs on which the cockroaches scamper are unique as well, and engineers are using their construction as the basis of robotic legs. Cockroaches have six legs with at least three knees on each leg; these legs allow the cockroach to make up to 25 rotations in a second – the highest known in any animal. The sensory spines on their appendages help them sense objects around them and move across very difficult terrain. Their hairy legs – and especially the tiny hairs called 'cerci' on two posterior appendages that we saw in Chapter 4 – also help cockroaches to be especially sensitive to the airflow around them.

And finally, though most of us would rather not know this, cockroaches do fly on occasion – especially during mating season. What isn't widely known is that cockroaches will come down from the ceiling to nibble on our wrists in the middle of a warm, humid night as well.

Like cockroaches, coyotes can quickly find and adapt to new environments. They can eat a variety of foods and also try to prowl around at times when possible predators such as the cougar are scarce. These canines have an excellent communication system – usually barks and howls – but can also set up and follow pheromones and other chemical signalling systems. They are equally aware of and in tune with various seismic vibrations.

Around ten o'clock one hot summer night, a sound like a marauding group of teenagers sprang up near Karen's house located on the edge of Washington's Rock Creek Park. Then the sound changed to that of a strange yelping. Soon after came a siren and then the strange 'party' whooping started again – very loudly.

Coyote vocal bouts can include both howls and a number of high-pitched cacophonous barks, and some of the sounds emulate or react to real sirens. This repertoire of calls allows coyotes to coordinate their moves and defences against intruders. Such is their astounding ability to use different intonations, that all the noise Karen had heard could actually have emanated from one lone coyote.

There are many ways in which coyotes have learned to trick their enemies. They have a special sensitivity to other animals and know how to use or outsmart them. For example, coyotes often hunt with badgers and grab their prey. They also move into badger or other animal burrows at times, if they don't build their own dens.

The adaptability and cunning of coyotes provide them with numerous advantages. They are especially good at figuring out the habits of humans, and are certainly more successful living around people than are their wolf cousins. And unlike wolves, coyotes will tolerate the lactating females in their pack. Sometimes coyotes will mate with domestic dogs or with wolves – although they mate with wolves less frequently because wolves just don't like coyotes. (The New England coyotes are thought to be part wolf.)

The range of hearing of coyotes is broader than that of wolves and dogs. Coyotes' upper frequency of hearing is 80 kHz, compared to 60 kHz for domestic dogs, for instance. And coyotes have sweat glands on their paw pads like dogs do – but wolves don't. Coyotes can out-race and out-jump their canine cousins: they can run up to 43 mph (69 km per hour) and jump over 13 feet (4 metres). And while coyotes function as pack members, it's not unusual for a lone coyote to travel over 100 miles (180 km) to scout for new territory or for other survival reasons. So the 'wily coyote' is a survival master. While it doesn't necessarily have unique survival attributes or behaviour, a coyote does more – more often and better – to live a healthy life and survive possible disasters.

Earthquakes: animal perceptions and alarms

Humans have the ability to be more alert to possible disasters than we often are. Our past experiences and where we live affect our alertness and the signals to which we might pay attention. For example, Californians are much more alert to ground vibrations than are those who live in other areas. Residents of London, New York and other cities might be more vigilant about possible terrorist threats than most rural residents. Others are more savvy about the smell of oncoming snow or the type of clouds and air composition that predict rain, while yet others know much about the precise feel of different types of wind activity. We are taught to listen a bit more to certain sounds or notice relevant visual cues to protect ourselves. Experience, training and scientific knowledge also help. Yet we can all benefit by understanding more about the perceptions and reactions of many animal species to natural occurrences. Our willingness to observe more carefully and learn about others can be life-saving.

Florida researchers have shown that sharks seem to leave an area that is being threatened by a hurricane days before it arrives, most probably due to their ability to detect subtle changes in air pressure. Screeching birds, nervous cats jumping out of windows, rodents running out of their holes, dogs barking, fish leaving the coastal area, and bees swarming in a seemingly rapid and purposeful manner are all examples of pre-earthquake animal behaviours that have been observed worldwide.

We introduced this book with the remarkable quote about the survival of numerous types of animals during the devastating earthquake and tsunami of 2004. Many of the extraordinary sensory abilities of animals that we looked at in Chapters 1 to 4 are reflected in the changes in sensory stimuli created by different types of earthquakes and tsunamis. There are myriad examples of signals that animals could have picked up to warn them of such an impending natural disaster.

Among the signals that could alert animals to an approaching earthquake are ground tilting, hygroception (awareness of humidity), changes in air pressure, and electric and magnetic, as well as mechanical, vibratory changes. According to Caltech scientist Joseph Kirschvink, these signals usually come before the earth really starts shaking. Although humans probably can't detect the tilting of the earth, many animals have much keener vestibular (inner ear and other balancing) systems. Such systems, for example, are much better developed in rodents that live underground in comparison to those that live above ground. While gravity is felt in an underground tunnel, there are no visual cues like the sky or plants to distinguish up and down. It's thought that homing pigeons have a sophisticated vertical sensitivity as part of their navigation system and that they too are naturally aware of ground tilting.

Humidity in particular is a key indicator of not just earthquakes but tsunamis as well. In addition to changes in groundwater level, moisture escapes upward into the atmosphere, meaning that the humid conditions can be detected by creatures living both above and below ground. Most if not all animals – including humans – are sensitive to changes in humidity. Some are especially sensitive, and spiders and some insects have special receptor cells with hygroscopic (having the tendency to readily take up and retain moisture) hair-like structures that detect humidity and temperature fluctuations. Shifts in air pressure can be an indicator of air escaping from underground cavities as well as storms gathering in the sky. Just as sharks have been found to leave an area before hurricanes arrive, the change in atmospheric pressure is felt by most animals. An example is the extreme sensitivity of the cerci of cockroaches to airflow.

In Chapter 1 we discussed how common it is for animals to detect electric and magnetic fields as well as mechanical and other vibrations. These may be given off weeks before as well as during the actual earthquake. Electromagnetic shifts are constantly taking place under and on the surface of the earth as rocks, lava and underground water move and collide with one another – even more so as ground pressures amplify. Electric charges can sometimes erupt, similar to the spark created when two pieces of flint or other rocks are rubbed together.

More subtle build-ups of electromagnetic fields can occur with two different kinds of rocks lying next to each other – a discontinuity between two types of rocks can create a great deal of energy under certain circumstances.* In addition, materials along a fault-line – which is especially vulnerable to extreme shifts – can have varying relative charges and therefore magnetic movement. Chemistry changes along a fault plane where there can be boiling water

* This discontinuity can result in what is referred to as second-order seismic waves in contrast to the stronger first-order waves that are called 'P' and 'S' waves (see page 120).

– hydrothermal veins where steam and geysers vent the pressure build-up. Such large as well as subtle movements and build-ups of energy contribute to the constant electromagnetic and energy changes that animals can detect.

New instrumentation has led to the development of the field of micro-tectonics, in which movements of rocks can be recorded at a very minute scale, right down to the level of atoms. One scientist has suggested that an earthworm might be able to detect a micro-tectonic fracture due to the pressure build-ups and swelling that precede macro-faulting and eventually mega-fault generation and movement in the form of earthquakes and subsequent tsunamis.

Seismic vibrations can travel great distances, and as we've seen in earlier chapters, many animals are equipped to pick up vibrations. A number of rodents such as California kangaroo rats use a low-frequency seismic 'foot drumming' to send messages between burrows and to notify predatory snakes that they have been discovered. These same 'drummers', as well as their snake predators, would be able to pick up precursor and earthquake vibrations. So could rabbits that 'thump' the ground to communicate. As we saw in Chapter 2, elephants are exquisitely sensitive to vibration, with pressure-sensitive nerve endings capable of detecting infrasonic (inaudible) vibrations in their trunks and feet. Sexually receptive female elephants send signals to males – deep rumbling alerts that carry more than 2 miles (3.25 km) in the air and more than three times that distance under the ground. Elephants can also sense the seismic vibrations from stampedes hundreds of miles away. It's not unreasonable to think that elephants that pick up earthquake and tsunami signals also send out warnings to other animals, and the same might be true of worms and many other inhabitants of land or water.

Earthquakes elicit a variety of sometimes contradictory reactions from animals because each quake is unique. There is tremendous geological variation along and within the earth. The pressures and physical components are different each day because of such things as climate, movement of subterranean liquids, and even man-made changes. For example, in Switzerland, the drilling of geothermal holes and the activity surrounding the drilling had to be curtailed because some small earthquakes – and one moderate one – ensued. Scientists are equally concerned about the possible effects of oil drilling in some areas.

In addition to these variations, there are at least four types of seismic waves that can cause and accompany tremors – two closer to the earth's surface and two types in the body of the earth – and certain animals are probably more sensitive to some rather than other waves.

The two surface waves, named for scientists, are Rayleigh waves and Love waves. Rayleigh waves usually cause the ground to roll rather than shake. Slower than body waves, they travel in ripples, as on water, and are responsible for the undulating of parked cars like waves on an ocean. Love waves travel slightly faster and tend to cause a horizontal shearing of the ground. As we saw in Chapter 2, underground burrowers like the mole rat seem to be particularly sensitive to both Rayleigh and Love waves.

'P' and 'S' waves are located much deeper under ground and they don't travel in straight lines, nor at constant speeds. P waves are the fastest kind of seismic waves, moving at many kilometres per second. They push and pull through the rocks like sound waves push and pull and create compression in the air. Unlike S waves, P waves can travel through liquids as well. Researchers have found that some animals can hear P waves moving through the liquids and solids of the earth and that this probably acts as a warning of the more serious tremors to follow; most humans may just feel a thump or a bump. Whereas the P wave is a compression wave, the S wave is a shearing wave. Although slower and unable to travel through liquid, the amplitude of the S wave is several times larger than that of the P wave, and it is this that causes much of an earthquake's destruction.

The variation in conditions within the earth and the types and timing of earthquake waves indicate that a large number and variety of signals can be detected from weeks to seconds before an earthquake and any resulting disasters such as volcanoes or tsunamis.

They are so finely tuned to the nature of the world around them that animals usually have time to escape the devastating effects of earthquakes, since it's only very close to the epicentre that the shaking starts without much of a warning time. Animals living about 10 kilometres from the epicentre, for instance, have seconds to minutes after detecting the P wave to escape the wreckage of the strong follow-up S waves. And since there are often various types of small shocks and waves of lesser strengths given off days and even weeks before the actual large quake, animals may have learned to heed those seismic forewarnings as well. Sometimes humans can also pick up the signals of an approaching earthquake, but our awareness of them is not nearly so honed as in other species. We would do well to heed the alarm signals of animals.

6

From Frogcicles to Dreamstates

Sleep that knits up the ravelled sleave of care,
The death of each day's life, sore labour's bath,
Balm of hurt minds, great nature's second course,
Chief nourisher in life's feast.

William Shakespeare, *Macbeth* (c. 1605)

Rock Creek Park in Washington, DC is home to a variety of different animal species monitored by wildlife biologist Ken Ferebee. Since 2002, several box turtles in this vast expanse of woods have been fitted with radio transmitters (antennae and all) to determine their habits and the reason behind their substantial decline in population. Turtle #72 is often around Karen's property and hibernates nearby under six inches of leaves and loose dirt.

Although in some years the weather vacillates from chilly to springlike and then back again, this doesn't seem to confuse the hibernating animals. This is because for many creatures, hibernation is not actually set by air temperature but by the angle of the sun and the corresponding length of day. Of course there are some exceptions among Rock Creek's residents – most notably in the form of tiny spring peeper frogs that actually freeze during winter, which obviously would be more affected by temperature change.

Either way, though, for more or less all of the winter, the hibernating animals will remain locked away from the harsh conditions that characterise the time of year. It's easy to wonder what – if anything – our turtle might be experiencing. Do turtles dream during these cold winter months? There's certainly ongoing research that suggests that animals don't simply switch off for the winter and back on for spring. Instead, many animals that hibernate will actually wake up once a week to go to sleep. Yes, you read that correctly.

It appears that deep hibernation and the more shallow hibernation state called 'torpor' are not substitutes for actual sleep, which most if not all animals seem to need. Hibernating golden-mantled ground squirrels (*Spermophilus lateralis*) raise their body temperatures every five or seven days. And for approximately 24 hours they go through slow-wave sleep. Then they cool back down and return to their hibernating state. So even during deep hibernation, animals seem to need a good dose of sleep, and scientists are gathering data to show that for some species, dreaming takes place during the sleep period.

Dead or alive? Don't ask a bear

You may think you've come across a dead animal. It doesn't move; in fact, it can be touched and moved around. The body is cold. And there's no noticeable sign of a heartbeat or of breathing. But in the midst of winter, there's a good chance this 'dead' animal is hibernating.

Many animals go into hibernation in cold weather to conserve energy at a time when their food supply is scarce or non-existent. While some hibernators actually freeze (the frogcicles in the pond), most go into low body temperatures that match the outside climate. Generally, body temperatures dip from 98°F (37°C) to around 30°F (–1°C). Rodents with incisors that grow and have to be 'trimmed' by the hard food they eat stop growing these front teeth during deep hibernation. Heartbeats reduce from as high as 80 beats per minute to four or five beats. Many mammals grow around three times as much fur this season. And they eat like crazy prior to their long winter respite, storing two distinctive kinds of fat: the regular white fat; and a denser brown fat that surrounds essential organs like the brain, heart and lungs. When the animals do come out of hibernation, this brown fat provides a quick burst of energy to warm them and get them moving and thinking quickly.

There are different levels of hibernation, from the frozen amphibians and reptiles to mammals that go in and out of hibernation (so don't assume that grizzly won't wake up if you bother him), to animals that go into torpor, the mild kind of hibernation. And there are hot weather hibernators. Just as animals hibernate to survive the extreme cold, others hibernate to survive extreme heat and the depletion of food and water. If you lived in a desert with temperatures that went as high as 134°F (57°C), you'd want to take a siesta, too. That's called estivation. Up until recently there had been no scientific evidence of hibernation or estivation among primates. However, researchers have now found such a primate on the island of Madagascar. In this case, the primate is a lemur that goes into a type of estivation when the climate is too hot and the food supply too sparse.

Bears and other mammals go through a variety of hibernation states depending on climatic conditions, amount of food and water, as well as mating demands. For example, brown bears are known to eat as much as 90 pounds (40 kg) of food each day to double their weight as they go into hibernation. Usually they dig dens in a hillside. And female bears are often pregnant as they go into hibernation, during which they may go into a lighter hibernative state to give birth to one or more cubs and nurse them until spring. Another mammal, the Ozark big-eared bat, goes into hibernation in its own style. Those gigantic ears, which are usually erect, coil up tightly like a ram's horn, conserving valuable energy since the outspread ears lose a great deal of heat.

Some lesser known examples of hibernation occur around us constantly – animals that use up so much energy they go dormant for a few hours each day. If you've ever wondered how a hummingbird could possibly keep up that beating of wings all day long – it probably doesn't. It conks out for a few hours to refuel. Other animals, like hamsters, can go into and back out of torpor daily. Yet they still need a good dose of old-fashioned sleep.

While it may seem perplexing that hibernating animals would wake up once a week to sleep for about 24 hours, Stanford neuroscientist Craig Heller documented this very fact in research showing that after stages of torpor, certain animals still need to go into deep sleep. After periods of hibernation and torpor, the nerve cell fibres (dendrites) have been shown to shrink as a necessary protein breaks away from the cell. But unlike brain deterioration or neuropathology – where this protein, which is used in brain development, learning and memory, breaks down – following heavy or light bouts of hibernation, it is remobilised and the brain dendrites grow again during sleep. However, if this seems complex, some animals have an even more extreme process of shutting down and coming back to life.

The peeping frogcicle

It's no bigger than a fingernail, 'peeps' lustily in early spring, climbs trees and can freeze into an ice cube in winter. It's a tiny tree frog known as a spring peeper (*Pseudacris crucifer*) – a male spring peeper, to be exact. And while we're thinking in exact terms, although this frog is able to climb high up in trees, it usually scales low plants (up to about three feet in height) with the help of sticky toe pads. Just as winter starts to recede, thousands of male peeper frogs excitedly hop like crickets in tall grasses around ponds, streams and marshes – looking for mates. A male spring peeper nearly doubles in size as his vocal sac enlarges with his shrill and persistent peeping. Displaying such action and such determination, it's hard to believe that he has only just thawed out from a frozen state; that just hours before, his brain, his heart and even his spry and sure legs were solid ice.

The spring peepers and their cousins the wood frogs (*Rana sylvatica*) possess what scientists call 'freeze tolerance'. When outside temperatures go way down, two thirds of the frog's body water freezes. Their hearts and brains stop functioning. Body temperature plunges to between 21° and 30°F (–6° and –1°C). Such freeze tolerance allows the wood frogs to live as far north as the Arctic Circle and as far south as Georgia in the United States. They are able to survive the harsh cold because of a natural antifreeze.

This organic antifreeze keeps the frog cells from dehydrating too severely during freezing. In this fascinating process some of the water deep inside the cells remains liquid. Glucose made by the frog's liver keeps the tissue freezing point low, similar to the way ammonia lowers the temperature of a car's windscreen wiper fluid. This prevents damage caused by the cell shrinkage that occurs during freezing. When the weather warms, the tiny frog thaws from the inside out. Within a few hours, the heart thaws and restarts. Then the brain. Finally the limbs thaw. And away he hops to find a mate.

Other forms of wildlife freeze and unfreeze as well. A number of insect species, several other north American frog species, one type of European lizard, and a few North American turtles are known to endure deep freezing. A number of Asian animal species are thought to do the same. However, no freeze-tolerant mammals or fish have been discovered as yet.

One type of fish does seem to come close to freezing with its special version of hibernation. During the dark winter, the Antarctic cod get down on the sea floor, slow their heart rates and eat very little. Not only is the Antarctic winter temperature extremely cold, but the lack of sunlight makes it difficult for the cod to see their prey. Kevin Fraser, a marine biologist from the British Antarctic Survey, says that in the winter these cod become comatose. But about once a week, they seem to wake up and swim around for a few hours. Scientists don't yet know the precise purpose of this, but could it be the case that the fish are actually catching up on a little sleep? The Antarctic cod study's co-author, Hamish Campbell from the University of Queensland, Australia, says that many freshwater fish do become dormant in the winter's lower temperature, but the Antarctic cod's experience is much more extreme.

Some scientists who are focusing their attention on the various degrees of hibernation, and especially the freeze-tolerant abilities of some species, hope to bring medical benefits to humans. For example, a few years ago a professional American football player suffered a spinal cord injury in a game that left him paralysed. Quick intervention by lowering his body temperature is credited with saving him and allowing him to regain the use of his arms and legs. Not only are we learning about the benefits of using freezing in brain and spinal cord injuries or organ transplants, but we also want to know how to help stroke and heart attack victims. We are trying to learn what happens in an animal (like the frozen frog) that allows the heart to stop and go into cardiac arrest, then restart (reboot, if you will) and function in a normal and healthy fashion.

Catching Zzzzs

Horses sleep standing up. Cows have been known to keep their eyes open while napping. A certain type of baboon stands on tiptoe high in a tree to sleep. Dolphins and southern fur seals sleep with half a brain at a time. And the common swift can sleep while flying. Other birds sleep in groups, with the birds at the edge literally 'keeping an eye open', most likely for possible predators, and the opposite side of the brain awake. Reptiles, amphibians and fish all sleep. Jellyfish do, too; about three or four in the afternoon, they mosey on down to the ocean bottom and sit there until the sun comes up the next morning. Fruit flies sleep as well; Washington University researchers were some of the first to determine this. They had a group of flies that they thought had died – but when they tapped on the glass of the container the flies slowly woke up. It had been nap time.

A sufficient amount of sleep is essential for all animals. Sleep-deprived rats and flies die more quickly than those deprived of food. Lack of sleep significantly affects our ability to focus, our judgement, and our moods. Not only are humans grouchy when we don't get enough sleep, it has been shown that many heart attacks occur because of sleep problems. Other acute and chronic health problems are also a result of too little sleep. Adequate sleep is necessary for growth and for the maintenance of all of our bodily functions. We need sleep for learning and memory consolidation.

Dr William Dement's groundbreaking book, *The Promise of Sleep*, tells of the decades of research that confirm the 'vital connection between health, happiness, and a good night's sleep'. While adult humans need a roughly eight-hour chunk of sleep a night, teenagers require 9½ or more hours to facilitate their growth and brain development; younger children need more, and tiny babies sleep most of the time – although it doesn't seem that way to their exhausted parents. Other animals have different requirements. Horses, giraffes and elephants, for example, sleep only two to four hours a day, while opossums and most species of bats sleep up to twenty hours in a 24-hour cycle. While rats sleep in spurts at different times of the day and night, clocking in a sum total of ten to twelve hours in a 24-hour period, human circadian* rhythms dictate the need for a good block of sleep at night (preferably in a dark environment). Recently, scientists have used sophisticated technology to prove that our cells synchronise their energy systems to surrounding light and dark – the basis of circadian rhythms. There are also ultradian rhythms that occur approximately every hour and a half and are accompanied by bouts of greater or lesser energy and different brain processes and frequencies, as well as circannual rhythms that are more than 24 hours. The field of chronobiology that encompasses the study of these rhythms has attracted a number of sleep researchers who are particularly interested in these periodic influences. The University of Minnesota's Franz Halberg, a pioneer in the field, has postulated that environmental influences on our bodily rhythms can under some circumstances even change the DNA of an animal. It's another example of how recent scientific discoveries are challenging the old genetic

---

* The word 'circadian' comes from the Latin *circa* (about) and *dia* (day).

assumption that we are born with a certain DNA map that dictates what our body does for the rest of our lives.

Elsewhere in the animal kingdom, along with swifts, other migrating birds are thought to go to sleep on one side of their brain at a time while they keep flying. Indeed, all birds cut down on their sleep during migration – sometimes by two thirds – but they make up for it when they finally reach their winter home.

Fish have their own variety of sleep habits. Many have active and inactive periods. Some sharks, such as the nurse shark, have been observed resting motionless on the sea floor. Six-foot-long white tip reef sharks that are found among coral reefs throughout warmer oceans have been photographed sleeping stacked one on top of the other in caves during the day; a daily slumber party to prepare for the night hunt. Others like streamlined sharks and hammerheads tend to keep moving during sleep in order to breathe. For them, movement is necessary to push the oxygen-bearing water over their gills, as their fins are too flabby to do this.

Many of us think we have an elaborate bed-time preparation routine, but ours probably doesn't compare to the complex envelopment that a certain fish must go through. The brightly coloured parrot fish (*Scarus iseri*), with teeth fused together to resemble a parrot's beak, has an especially interesting pre-sleep ritual. The parrot fish prepares for nightly sleep by spewing out slimy mucus from a gland in its head and wrapping itself in a mucus envelope. This slippery cocoon forms in 30 to 60 minutes and hides the parrot fish from view. Predators are also deterred because this special 'sleep wrap' smells and tastes terrible.

Ready for flight

Since they don't have threatening teeth, body armour, or a noxious odour to spray on their enemies as skunks do (or wrap around them as parrot fish do), horses have to be ready to gallop away – fast – to escape any sudden danger. Horses are flight animals.

To achieve this balancing act, horses are endowed with sophisticated leg and hip systems to facilitate sleep while standing. What veterinarians call a 'stay apparatus' prevents movement of leg and upper joints during sleep so a horse's legs don't collapse, and a locking mechanism of the knee and surrounding areas allows the horse to put weight on one hind or front leg while resting the other side. All the horse has to do is rotate its hips and legs in a certain manner to set in place the stay apparatus and locking mechanism, thereby creating an effective sleep 'stance'. So if a horse has its head down, bottom lip drooping and one hip sagging, it's probably taking a nap.

There are times when a horse might just decide to lie down for a while to sleep, and some horses do this often. Generally, however, horses and close relatives like donkeys and mules stay prepared for flight, even while they sleep. More distant relatives like zebras also sleep standing up. Those in the wild usually don't go to sleep at all unless they are surrounded by other alert zebras, or unless other types of vulnerable animals like them are around to warn of danger. Animal sentries must watch for lurking lionesses, cheetahs or other predators. Such predators can afford deeper bouts of sleep than their potential prey. And because the male lion isn't usually the hunter, he can really enjoy a long, leisurely sleep.

Deer, such as the white-tail, sleep lying down while carefully camouflaged in a bed of vegetation. They have been reported to sleep with their eyes open or closed, but always vigilant at some level. Even some humans in wild territory have to remain vigilant. When Karen spent some time in the Kalahari Desert learning about the customs of the !Kung peoples, she was told that they tended to sleep in short spurts, preferably with at least one group member awake and on the alert for any possible danger. Nomadic Somalis that she knew – and probably many other wilderness dwellers throughout history – made sure they had a campfire going throughout the night to fend off wild animals. In more familiar circumstances, many of us rely on such devices as burglar alarms and smoke detectors to 'stand watch' while we sleep.

Animals that dream and what they dream about

And perchance to dream? Humans average four dreams a night, but usually remember only about one a week. Most of us believe that animals dream. We've watched our dog or cat sniff or yelp softly as they slept. We've seen their legs moving as if in a run while the pets were obviously sound asleep. Scientists have also known for a long time that animals dream. But Matthew Wilson and his colleagues at Massachusetts Institute of Technology were the first team of scientists to find out what animals are dreaming about, and that their dreams are complex. Wilson and his graduate student, Kenway Louie, gave rats a food reward for running on a circular track – and tested what brain cells were activated during this task. Later, as the rats slept, the MIT researchers found that the same neurons fired during REM (rapid eye movement) sleep – the sleep stage in which most dreams occur – as had fired during the earlier waking activity on the track, and were even able to tell what part of the track or maze each rat was on during certain times in the dream testing. The rats were dreaming about what they had done that day and putting it into long-term memory. With the aid of their dreamstate, they were learning.

Before this research, only a few animals such as chimpanzees and dolphins were thought to be able to recall and evaluate sequences of events after they occurred. This study proves that rats and probably many more types of animals can remember and re-evaluate events previously experienced. As complex as these dreams were, Wilson suspects that non-lab animals have even more complicated dreams, because they lead more interesting lives. His rats were quite sheltered. Furthermore, Wilson and Louie found that each rat had its own signature pattern of brain waves – just as Karl Pribram of Stanford and Georgetown universities has found with human subjects. This signature pattern of individual animal brain waves is still not known by much of the scientific community, although it has wide implications.

So, from frogcicles to dreamstates. Various hibernation levels, sleep, and dreams are just some of the many animal secrets that humans are beginning to probe and find endlessly fascinating.

7

Animal Marathons by Land and Sea

The journey of a thousand miles begins with one step.
Lao Tzu (604–531 BCE)

Mice giggle,* and they may have good reason to do so when we naive humans drive the little field mouse we have captured in our house a short distance away, thinking it won't return. This chapter will look at the distances that animals travel, and how they know where they are and where they're going. Scientists have shown that, like human travellers, other species use the sun, moon and stars as guides; without the sophisticated technology that we may need, they know how to tap into 'radio' frequencies as well as the magnetic fields of the earth. The monarch butterfly can travel thousands of miles to a small location near Mexico City using what's known as a 'timing compensated

* We will look in closer detail at *why* mice giggle in the next chapter.

sun compass' located in a brain no larger than the head of a pin. This compass is continuously calibrated by an internal circadian clock. Surprising to the scientific community, the monarch's complex navigational system is more like that of a mouse than of other insects.

Among migrating birds, the arctic tern is the ultimate long-distance flier. Every year it flies from the North Pole to the South Pole – and back again. And large land animals also travel great distances – always spectacular is the yearly migration of millions of wildebeest across the vast Serengeti Plain of East Africa. Scientists are discovering more and more interesting and spectacular ways in which animals travel.

Among some of the more unusual lone fliers is a tiny migrant that just takes off on a thread and a breeze before travelling for hundreds of miles.

The flying spider

Until recently, scientists couldn't explain how spiders could travel hundreds of miles over water to populate volcanic islands or visit ships far out in the ocean. Generally, spiders travel only a few metres. Yet some take off and 'fly' incredible distances. Former theories couldn't explain this anomaly until Andy Reynolds and colleagues at Rothamsted Research studied these tiny creatures and their amazing ways. By standing on tiptoe and casting out a single strand of silk, spiders are able to parachute and ride wind currents. While they may often intend only to hop a short flight on the silk, spiders, in their efforts to flee a predator or colonise new places, can get caught up in a turbulent airflow – for miles and miles and miles.

Turbulence feeds upon its own momentum, moving rather chaotically and then reorganising, or self-organising, itself to a type of temporary stability – then reorganising again. Vortices or swirling eddies are examples of this non-linear dynamic. Air currents offer even more leeway than water, and some air turbulence can be a part of a thermal column that creates its own way of sustaining the turbulence. The soaring bird of prey can soar because of thermal draughts formed by the sun's warming of the earth and the air rising when it becomes heated. While it has been known for years that all migrators take advantage of thermal columns, that seems to be only part of the story. Turbulence theory and the mathematical formulations comprising its actions can explain even further what happens to these thermal columns or draughts and to the fliers taking advantage of these phenomena. Meanwhile, the little spider that caught a 500-mile ride on a thread can rest a while as it settles down on a new island or a far-off ship.

Marathon fliers

Imagine flying over five million miles – not on a commercial jet, but with your own wings. That's what a certain Manx shearwater is estimated to have flown over its 50-year lifespan. (And it was still flying in 2008.) That bird probably holds the known lifetime long-distance record – combining flying strength with remarkable longevity. The Manx shearwater breeds on islands and coastal sites around Britain and Ireland. Skokholm and Skomer islands are known to have the largest concentration in the world of these birds; ringer studies from these two islands show that some of the shearwaters make their yearly 6–7,000-mile (10–12,000 km) migration to their winter quarters on the coast of southern Brazil and Argentina in less than a fortnight. As they soar gracefully overhead, their bodies change from black to white as they dip to show their black top parts, then rise to display a brilliant white underside. They are beautiful sights to behold, these members of the elite long-distance fliers club.

Another long-distance flier, the wandering albatross, has flown in our poetry for centuries and captured our imaginations as well. Its grand wingspan outstretches more than ten feet – greater than that of any other flying bird. And for the first ten years of its life, this wanderer rarely if ever makes landfall. Instead, it rests on undulating ocean waves and often follows fishing vessels to feed on discarded fish. Perhaps at times it encounters a wispy spider also navigating those vast and lonely seas. This impressive bird flies many thousands of miles across the southern hemisphere from its breeding grounds on islands near the Antarctic circle up to the equator, often circumnavigating the globe to find food. Researchers tracked one bird that flew an incredible 3,700 miles (6,600 km) in less than twelve days. When the wandering albatross is ready to mate, it returns to the tussock grass-covered islands of its youth to build a nest. There, both male and female take turns sitting on the lone egg and then tending the chick while the other flies off to get food.

Songbirds don't do marathon flights; they have prolonged stop-over sites on their migration routes. However, a group of these birds recently surprised Bridget Stutchbury and her team of researchers from York University in Toronto, who fitted them with little geolocator tracking systems. A sensor system, weighing a little more than 1 gram, was placed on each bird's back, right above the hips. (Yes, birds have hips.) This minute bird backpack was then secured with a little loop going around each leg. The songbirds travelled three times faster than predicted – some clocked more than 300 miles (500 km) a day. One purple martin, a member of the swallow family, took 43 days to get from Canada to Brazil during the autumn (because it stopped for a few weeks of R and R in the Yucatan), but in the spring it made the same trip back to its northern breeding colony in only thirteen days. The martin is famous in much of the US for its migrations and for sending scouts back north to check out nesting sites in the spring a few days before the main flock of martins returns.

Some swallows, close relatives to martins, are legendary for coming back to their nesting grounds on exactly the same day each year. One particular group of swallows has become famous for its timely precision. As a teenager, Karen would go with friends to the old mission of San Juan Capistrano, south-west of Los Angeles, on 19 March to welcome the yearly arrival of 'the swallows of Capistrano'. The swallows return the same time every year to their roosts in the shelter of the old stone church. As with the martins and many migrating birds, the 'scout swallows' return a few days before the main flock flies in, early in the morning on St Joseph's day. There's an almost uncanny sense of timing too when these fliers leave again as the colder weather arrives. Every year on 23 October, the day of San Juan, the swallows of Capistrano leave for the winter. After circling the mission a few times, bidding farewell, they fly away in a swirling mass.

The monarch, one of the most beautiful of butterflies, is also one of the most amazing. Steven Epbert of the University of Massachusetts studied the migrating monarchs that can cover 3,000 miles (over 5,000 km) in their migrations* and found that they use the 'timing compensated sun compass' mentioned earlier. Epbert studied the circadian clock in the tiny brain of these butterflies that works closely with the sun compass information, which itself processes photo-sensitive information in the eye and neurocircuitry in the mid-brain. The calibration between the clock and compass is continuous. If the circadian clock were destroyed, the butterfly might still follow the sun, but couldn't compensate for the change in the sun's angle each day. The clock and the compass work together as what some scientists call a 'positive feedback loop' and other scientists call 'fast forward feedback' – information is given back and forth constantly between clock and compass and the flight forward is adjusted accordingly. One way that scientists were first able to get a detailed understanding of these amazing circadian clocks was based on their studies of the clock-genes of fruit flies and of mice. These genes create and interact with RNA (ribonucleic acid – a messenger and regulating molecule) and protein. We can only marvel at such intricate ability in exquisite creatures that swarm and travel in groups of thousands in a flying dance of unbelievable beauty and precision.

---

* Incredibly, in some specific cases, a single migration can span more than one generation.

Checking the scale and getting in shape

The beautiful monarchs don't have to worry about their weight. But many animals do. In another form of calculation and calibration, migration flocks have to be hardy enough to fly long distances without being too heavy to fly efficiently (or, frankly, without being too heavy to fly at all). Before taking its 500-mile (900 km) trip to Central America, the ruby-throated hummingbird feasts on insects, sweet nectar and tree sap. This tiny bird must gain at least 2 grams of fat – almost twice its usual weight – before making this non-stop flight.

In contrast, young swifts may go on a diet. Jonathan Wright and his colleagues at the Norwegian University of Science and Technology found that these young birds seem to gauge how much they weigh by doing push-ups with their wings. Then they decide if they need to lose some weight in order to fine-tune their bodies in preparation for making a non-stop migration over thousands of miles; they must fly continuously and, as we saw in Chapter 6, even while asleep, until they reach their breeding grounds.

Another fast-flying migrating bird, the bar-headed goose, has to be in tip-top condition as well to be able to fly over 50 miles an hour up to an altitude of 34,000 feet when crossing the Himalayas. Bar-headed geese can fly over 1,000 miles a day (1,600 km). In order to not only survive but fly rapidly at such high altitudes where the air is thin and the oxygen supply sparse, their red blood cells must be uniquely able to absorb oxygen more efficiently than the blood cells of any other birds. Not only that, but in order to get the lift they need, the ratio of wing surface to body weight must be greater than in other species. And humans worry about jogging a few miles.

Reducing drag and keeping track of the group

We can't refer to geese migrations without mentioning the research being done on their ingenious flight formations. On the open highway, many of us get quite nervous when we see a gigantic lorry dangerously tailgating an even bigger one. This is done to save on the costs of petrol and general mechanical operations. The first lorry goes into the wind and shields the second lorry, which benefits from this reduced drag. A large group of cyclists does the same.

Geese have always used this trick. That's one reason for flying in a V pattern. Scientists have found that in a V formation of 25 geese, each bird can achieve a reduction of induced drag of as much as 65 per cent, and can increase its range by up to 71 per cent. This allows the birds to fly long distances efficiently. All the birds but the first one are flying in the upwash from the wingtip vortices of the birds ahead. A little upwash helps the geese support their own weight in flight, in the same manner that gliders can climb or maintain their height indefinitely in rising air. (Remember those thermal draughts.)

The geese take turns flying first, and when the front goose is tired, it goes to the back. Those birds flying at the tips are rotated in a cyclical fashion as well. Many times the V formation is lopsided, a little longer on one side. This seems to help in two ways. First of all it conserves energy. Each bird flies slightly above the bird in front of it – taking advantage of the uplift provided by the vortices and small tornadoes caused by the wings and tail configurations of the bird ahead, and further reducing wind resistance. Secondly, this pattern makes it easy for birds to see and keep track of every bird in the group. Humans have taken note of this flying precision and used it to perfect their own flights. For many years, fighter pilots have used this formation learned from the geese. With increasing fuel prices and other financial difficulties, commercial airlines are now exploring how they can also benefit from these flight patterns.

Energy conservation is certainly a concern for one type of tiny bird. The hummingbird never seems to stop. It can flap its wings up to 80 times a second and has a heart that can beat over 1,200 times a minute. To keep this vigorous metabolic pace, this busy bird must feed constantly on sweet, energy-giving nectar. But as we saw in Chapter 6, at such a frantic pace the hummingbird must also stop, not only to nap, but to go into torpor – a light state of hibernation. When it does fly, the hummingbird's versatility can't be beaten. It can fly sideways, straight up and down, even briefly upside down, and of course it can also fly backwards. The only other birds observed flying backwards just a little on occasion are some flycatchers and warblers. The hummingbird's specialised wing shape, muscles, and metabolism allow it to hover in place like a living helicopter. The ability to hover makes all the rest of its unique flight manoeuvres possible. Humans are working to create tiny robots that use some of the rotations, back and forth, up and down, forwards and backwards, and hovering-in-place abilities of the hummingbird. Such little robots could be invaluable, for example, in finding and rescuing those in danger or in sophisticated surveillance strategies.

The great land migrations

When Karen lived in Africa, she had the awe-inspiring experience of camping in a small tent for two weeks in the Serengeti Plain in the midst of the annual migration of millions of wildebeest. Wildebeest travel more than a thousand miles in huge herds along with thousands of zebra, gazelle, and elands in order to find greener pastures during dry seasons in Kenya and Tanzania. The migration takes a circular route around Serengeti National Park to find the right food, water and minerals – especially phosphorous that is vital to the animals' health.

While Karen and her friends had to be cautious about lions and snakes, the time of the great migration is supposed to be the safest for human camping. That's because the predator lions that follow the migration have an abundant food supply around them, and really don't care for human meat. Humans taste terrible. So we're not the preferred dinner of a lioness, and are vulnerable only to a sick or very old lion who can't get a decent meal otherwise.

Hyenas also follow the migrations and gang up at night to fight over and finish the meals that lions might have started. But hyenas are necessary, as are other scavengers like vultures – someone has to clean up. And all animals have to be cautious of the predatory crocodiles that lie in wait near the rivers. Unfortunately, this integrated group of animals that forms the most impressive of all land migrations is threatened more and more each year by the effects of climate change, urban development, and violent poaching gangs. The group of over 5 million estimated migrating wildebeest when Karen first experienced the migration

30 years earlier had decreased to a little over a million when she revisited East Africa in 2004.

One migrating mammal has received bad press over the years. Many stories have been told about how lemmings commit mass suicide by jumping off cliffs. In fact, it's mass panic that can have these disastrous results. For example, Norway lemmings sporadically migrate every few years when their populations get too large to survive comfortably in certain regions. Typically these animals live in tunnels under the snow during the winter, then move towards mountainous meadows or forests in the spring. However, an especially mild winter results in an earlier than normal spring and a later autumn. Food becomes so abundant and the weather so pleasant that large litters result. By summer, the lemming population may be so large that mass migrations are triggered. During such migrations, the extraordinarily large numbers of lemmings may encounter obstacles like boulders, rivers or cliffs that force them into bottlenecks. Then in panic these little animals may go en masse over high cliffs or into rapidly moving waters. Even though the reason for the lemmings' demise is not as the legends have supposed, the result is just as destructive.

Migrations by sea

Of all migrating mammals, the whale travels the farthest. The humpback whales of the northern hemisphere travel from their summer feeding grounds in the north Atlantic and Pacific oceans to their winter breeding bases in Hawaii, Japan, Mexico, the Caribbean and West Africa. The southern hemisphere humpbacks spend their summers in the clear, cold waters off Antarctica and head to warmer winter waters in Australia, southern regions of Africa, or South America. They can travel over 3,000 miles (5,000 km) on the first half of their migratory cycle. And as they go to warmer waters near the equator, the male humpback males escort the females to breeding grounds – serenading them along the way with complex songs. Any competitors in this courting ritual are warned away by aggressive tail-slapping by an irritated male. After it's born in the breeding areas, the calf swims the thousands of miles back with its mother to colder waters, and they may stay together for over a year. If the humpbacks sense danger at any time, they are known to give out strange-sounding sets of grunts interspersed with vigorous bubble-blowing.

Whale-watching humans sometimes witness these humpback alarm signals or calls from other whale species. They may even be entertained by the beautiful courting songs the males sing. And on an especially good day, whale-watchers may be fortunate to observe a whale spy-hopping. In spy-hopping, the whale suddenly lunges vertically out of the water and spins around, looking for useful landmarks on the long migratory journey. Millions of people witness the whales' spectacular yearly migrations. Many beach-dwellers keep binoculars nearby during 'whale season' to spot the huge swimmers at a moment's notice. Others go out on boats to be nearer the action. Becoming more knowledgeable and appreciative of the magnificence of wildlife can be a double-edged sword, though. For example, the sounds of boats and human onlookers have been found to disturb the whales and interrupt their mating and sleep patterns. In our eagerness to participate, humans often interfere with and can even harm the life-sustaining rhythms of those we seek to understand.

Those of us who have been pregnant may have had to decide at a certain point whether we were up to taking a long-distance commercial air flight. Pregnant green turtles (*Chelonia mydas*), however, swim over 1,000 kilometres in the course of a typical pregnancy. Although these turtles have dark brown shells that are white underneath, it's the distinctive green fat under their shells that gives them their name. These mothers-to-be swim from their feeding grounds in Brazil back to their birthplace of Ascension Island in the middle of the south Atlantic ocean. Upon their arrival, the pregnant turtles dig out nests on the sandy beaches, lay their eggs, then eventually head back to Brazil. A mother turtle can lay clutches of 60–150 eggs at a time, and up to eleven clutches within the space of a few days. The babies all hatch at the same time, about a month after the eggs are laid. Whether the baby is male or female is temperature-dependent. If the temperature of the egg is lower than 28.5°C, the turtle will be a male. If the temperature is greater than 30.3°C, the turtle will be a female. Soon after birth, the young scurry back to sea.

While whales and pregnant turtles are swimming above, hordes of spiny lobsters might be marching below on the sea floor. These brown-grey lobsters with yellow spots take off on their journey across the sandy bottom of the Caribbean Sea in October and November. Autumn storms often create so much turbulence and cold temperatures in the shallow waters of the Caribbean that these spiny creatures will travel up to 40 miles (60 km) to warmer ocean channels. They 'ride' the turbulence as a soaring bird might, but without leaving the sea-bed. They usually march 'in line', with each lobster using its antennae and front legs to keep in touch with the one in front. As with the truckers, cyclists and V pattern-flying geese, the in-line marching formation of the spiny lobsters also reduces drag. During this annual migration prompted by stormy seas, tens of thousands of these migrants can often be seen off the west coast of Bimini, in the Bahamas, moving southwards day and night for several days in oriented queues of up to 65 individuals.

There seem to be a number of common factors in migratory and other group behaviour – flocking, swarming, and schooling or shoaling of fish – as Nicholas Makris and other scientists have concluded, and as we mentioned in Chapter 2. A certain number of individuals within a certain area can initiate a migration, as can climatic conditions – especially the temperature and angle of the sun. A restlessness among the migrators will generally have been observed weeks before their big trip as they prepare in many ways. Population increase initiates various forms of group behaviour as well, for reasons other than annual migration to and from specific areas. An interesting biochemical factor of preparation has been noted recently by scientists studying the swarming behaviour of locusts, and it will be fascinating to see if further research bears out the same neurotransmitter change in other species. When they are not swarming, locusts lead relatively independent lives, and in fact they are generally repulsed by rather than attracted to other locusts – other than for mating, we might assume. However, some weeks before they start swarming, the locusts change or become brighter in colour and have greatly increased amounts of the neurotransmitter serotonin in their systems.

(Serotonin levels have been shown to be much lower in depressed and anxious people. Too much of it, however, has been associated with schizophrenia and psychotic reactions.) A number of studies have been initiated to explore the role that neurotransmitters such as serotonin may play in animal gatherings and group movements. Scientists will look at the difference between biochemicals and population thresholds that bring animals together to live and to move in beneficial ways and the dangers of chemical deficiencies or excesses as well as over-population, and consequential 'bottlenecks' within a confined area – conditions, as the lemmings have warned us, that can create panic and destroy lives.

Other ways of getting around

They may not travel as far, but a number of animals have their own interesting ways of moving along.

A recently discovered, gorgeously coloured and patterned fish travels like no other fish – it hops. The *H. psychedelica* lives up to its name. This beautiful flat-faced fish with apricot and white swirling stripes and bright blue eyes that both face forward, as in humans, was found in the Ambon Island region of Indonesia in 2008. Ted Pietsch from the University of Washington was the first to publish a paper on this fish that bounces around under the surface of the ocean like a happy rubber ball. Pietsch also gave *psychedelica* its name. While other related species jet propel up from the sea floor as they go into swimming mode, the *psychedelica*, mobilised by jet forces coming from gills on each side of its body, just keeps hopping and bopping around.

A desert- and forest-dwelling arachnid of California, known as the solifuge or sun spider, does a wonderful cartwheel all the way down a hill. When the spider wants to escape its enemies, all it has to do is curl up in a ball and roll on down the nearest sand dune. Because this tiny creature is often mistaken for a minute wispy tumbleweed, most humans never realise the incredible acrobatic animal act they are witnessing. Ingo Rechenberg, a Berlin Technical University bionics engineer, found that the rolling Sahara spider (*Araneus rota*) also does an impressive cartwheel. The palm-sized arachnid gets a running start, then propels itself into a series of down-dune cartwheels at the rate of about 5 mph (9 km per hour).

There are also some interesting walkers in the animal world. Some only go sideways and others 'wave' along. Besides the side-winding snake, which isn't really a walker – but it does the best it can without legs – there's the well-known side-walking crab. Most crabs have the capacity to move laterally even as they move forwards. They go sideways because of the direction their legs bend, and this sideways technique can often confuse predators geared towards expecting more forward-moving locomotion.

Centipedes and millipedes, distant cousins of lobsters, crayfish and shrimp, are land arthropods. With so many legs, how do they keep from tripping over themselves? Centipedes can have over 100 legs. Millipedes, despite their name, usually have only up to 400 legs, except for one type that has an incredible 750. All these legs enable the partially burrowing centipedes and millipedes to successfully navigate loose dirt, leaf remnants, and other materials. Their tiny legs only move forward and never to the side, pushing powerfully through the debris. They move in waves – about twelve legs at a time pick up and then go down, with the following legs doing the same, all the way to the back; then starting again at the front.

★

In this chapter we have shown that there are many different ways in which animals get to where they would like to go, from kiting hundreds of miles over the ocean on a spider thread to moving hundreds of legs in coordinated waves. Locomotion – the act of moving from place to place – separates animals from plants. And it's thought by many neuroscientists such as Rodolfo Llinas of New York University that such goal-directed active movement, a biological property known as 'motricity', is a requirement for the development of the nervous system. In his book, *i of the vortex*, Llinas shows how motricity starts the processes that result in an animal brain. At the very least, it's interesting to move around and experience new places and to learn as we do.

Part III

Socialising

8

Wit, Wiles and Good Fun

C'est une grande habileté que de savoir cacher son habileté.
(The height of cleverness is to be able to conceal it.)
Duc de la Rochefoucald, *Reflexions ou Sentences et Maximes Morales* (1678)

Mice giggle when tickled by a human hand. They must enjoy the tickling, since they follow the person's hand around the cage seeking to be tickled again. It's well documented that many animals, humans included, babble, sing and talk using their own unique sounds. Humans also giggle, of course, and such are the shared characteristics between species that it's perhaps only logical that many other creatures should giggle as well. Whether they do so in delight, in fun or just because they're tickled is, as yet, unclear. Lab experimentation is the best, albeit contrived, way to begin to scientifically document our findings.

The first researcher to document emotions and laughter in animals was Jaak Panksepp, while studying vocalisation in rats. Ongoing studies show that rats make high-pitched squeaks when they play. In addition to Panksepp, neuroscientists like Sergio Pellis of the University of Lethbridge in Canada and Andrew Iwaniuk of Monash University in Melbourne assert that play is an important process in the development and maturation of the growing brain. Play has been shown to facilitate the growth of brain cells and connections in the area of the amygdala (a part of the brain that functions as a type of radar system and that processes emotions) and the dorsolateral prefrontal cortex (the executive area in the brain that facilitates judgement and decision-making). Fun and spontaneous, play is a way to learn about and rehearse life skills – a way to connect with and test the limits of others. Play is serious. It's an important way for children to learn about the world and for adults to passionately and creatively live their lives and continue to grow brain cells and neuronal connections.

In his book *Play*, physician Stuart Brown writes that 'play is a profound biological process [that] has evolved over eons in many animal species to promote survival. It shapes the brain and makes animals smarter and more adaptable. In higher animals, it fosters empathy and makes possible complex social groups. For us, play lies at the core of creativity and innovation.' Brown's definition of play is more expansive than many of us may have considered. It includes doing work we love to do, and seeing humour in virtually all situations, as well as aesthetic appreciation. 'Simply taking a moment to deeply inhale the air after a rainstorm or kick a pile of leaves can be a private moment of play.' We give play signals to others – a smile, an outreached hand, a hug, good-hearted teasing or joking, gestures of silliness.

Brown, along with other researchers, has studied play in a variety of species. In addition to mammals, birds and some reptiles, they found that octopuses play. These creatures have been known to playfully take the tops off jars, put the tops back on, and toss the jar around in the water for a while – in a relaxed and rather idiosyncratic way. Certain territorial fishes blow bubbles in a manner that appears to be play. Even ants play – at least according to ant expert E.O. Wilson, who believes that ants engage in play-fighting.

We all know when our pet cat or dog wants to play. Dogs bark, wag their tails, stretch down into a play bow. And as Stuart Brown describes, they 'perform a rapid panting akin to human laughter (chimps and rats have also been shown to perform this panting with vocalization when communicating excitement)'. Sometimes we even get down on the ground with our dogs in a play bow, then a mock sparring or wrestling. Play signals 'invite a safe, emotional connection, if even for an instant'.

Brown also writes about and includes photographs in his book of an incredible encounter between a dog and a very hungry wild bear:

> Hudson seemed to be a very dead dog. That's what musher Brian La Doone thought as he watched a twelve-hundred-pound polar bear quickstep across the snowfield, straight toward [his] sled dogs. … [J]udging from the appearance of this particular bear [La Doone] knew it had not eaten in months. … As the bear moved in, Hudson didn't bark or flee. Instead, he wagged his tail and bowed, a classic play signal.

> To La Doone's astonishment, the bear responded to the dog's invitation. Bear and sled dog began a playful romp in the snow, both opening their mouths without baring their teeth, with 'soft' eye contact and flattened hair instead of raised hackles – all signaling that each was not a threat.
>
> In retrospect, the play signals began even before the two came together. The bear approached Hudson in a loping way. His movements were curvilinear instead of aggressively straightforward. When predators stalk, they stare hard at their prey and sprint directly at it. The bear and the dog were exchanging play signals with these sorts of curving movements as the bear approached.
>
> The two wrestled and rolled around so energetically that at one point the bear had to lie down, belly up: a universal sign in the animal kingdom for a time-out. At another point during their romp, the bear paused to envelop Hudson in an affectionate embrace. After fifteen minutes, the bear wandered away, still hungry but seemingly sated by this much-needed dose of fun.
>
> Every night for a week, the polar bear and Hudson met for a playdate. Eventually, the ice on the bay thickened enough for the famished but entertained bear to return to his hunting grounds [to feed].

A very different example of enjoyment in the animal world is the popular video of bottlenose dolphins playing with toys of their own creation. Through a blowhole on the top of its head, a dolphin creates a bubble and then, with a flip of its head, forms a beautiful silver ring from the vortices of the bubble. Then the dolphin pokes its nose through this splendid toy, tosses it around, and plays.* Dolphins cavort and share their bubble rings. Eventually, a ring will start to collapse and the dolphin will nudge it a certain way so it breaks into tiny effervescent bubbles. This water artist creates a new silver bubble ring and the games go on. Theorists of mind marvel at the 'intentionality' of this game; these dolphins intentionally create the bubble ring, and they create it for a purpose – to have fun. Physicists and engineers appreciate the adeptness involved in making such rings – toys that don't rise away to the water's surface, but stay at a level for the dolphins to play with and enjoy.

There's also reason to believe that animals have a sense of humour. The African grey parrot has been known to make up its own jokes. And animals can go from playful to downright mischievous, as renowned neuroscientist Karl Pribram (Professor of Cognitive Neuroscience at Stanford and Georgetown universities and a pioneer in the study of primates) can attest in his many stories of the antics of the monkeys, chimpanzees and gorillas he studied for half a century.

In appreciating the wit and wiles of animals of many species, it's important to remember that 'wit' not only means humour, it means intelligence and a learned mind; in fact, 'wise' comes from the same Middle English/Anglo-Saxon root. Two different animal groups exemplify in their own ways witty behaviour in animals: parrots – the big brained avians – and non-human primates.

* http://www.youtube.com/watch?v=TMCf7SNUb-Q

The amazing abilities of African grey parrots

> 'Richard … Richard … RICHARD … Come here, Richard … Richard, please come over to see me … to see me … me … mumble, mumble … that bird … mumble … Richard … I'm lonesome, Richard … Come over here, please, Richard … mumble, mumble … Richard … Richard … WHERE'S JENNIFER, RICHARD?'

Sounds like a four-year-old scheming to get attention. But it's really a 22-year-old African grey parrot named Toby. Toby was getting plenty of attention from the scores of audience members in the auditorium that winter night. But Toby wanted Richard's attention. 'Richard' is noted neurologist and author Richard Restak, who was presenting the introductory lecture before a large group attending the Smithsonian's National Zoo in Washington, DC. (Jennifer is Richard's grown daughter who, along with his wife Carolyn, was in attendance.) Karen had accompanied Karl Pribram and his wife, novelist Katherine Neville, to a programme about grey parrots highlighting the recent book and work of Irene Pepperberg, who spent more than 30 years training and studying Alex, another African grey parrot. Richard Restak began the night with an overview of the human brain and that of other animals, particularly big-brained birds. On the other side of the stage on a perch was Richard's grey parrot, Toby, who was participating, unsolicited, in Restak's speech about how bright grey parrots are. Another parrot trained by Pepperberg was in a cage close to Toby. There seemed to be both parrot-bonding and competition going on between the two birds.

Only recently have 'bird-brains' been treated with respect in academic circles. Pepperberg writes in *Alex and Me* that when she started doing the research on Alex, the 'received scientific wisdom ... insisted that animals were little more than robotic automatons, mindlessly responding to stimuli in their environments'. Pribram had championed Pepperberg's groundbreaking research, and Alex, who died at the young age of 31 (parrots are known to live twenty more years), contributed much to the change in attitude of the public and scientific community. This remarkable African grey parrot had a vocabulary of more than 150 words. He knew the words for more than 50 objects, could identify at least seven different colours and five shapes, and understood concepts like bigger or smaller, same or different. Amazingly, he even grasped the abstract notion of 'zero'.

Alex was witty and assertive and had no trouble expressing his likes and dislikes. 'Wanna go back', he'd say whenever he wished to return to the solace of his cage. Or to eat, 'Want a banana'. And if someone handed him a grape instead, he'd hurl the grape back at them. Like a human child who has learned some of the patterns of language, Alex would compose his own words if he didn't know the name of a certain food, for example. He made up the word 'banerry' (a cross between banana and cherry) for an apple.

He loved to have his head tickled. And when Pepperberg tickled him, the delighted parrot would nuzzle his head to the side and blush pink. He amused himself by tearing paper into shreds, and enjoyed eating cake that he called 'yummy bread'. He would dance to The Mamas & The

Papas and seemed to like classical music as well. In fact, when he was agitated, only Haydn's cello concertos could soothe him. When Pepperberg brought in other grey parrots to teach, Alex would become very jealous and bossy. He'd shout out answers before the other parrots could, and when they did answer, he'd chide them with: 'You're wrong … say better.'

The Restak parrot, Toby, doesn't spend time in a university learning lab; instead he has been 'home-schooled'. African greys are very sociable birds and want to be around others constantly. Sometimes when Carolyn Restak is making dinner she may put Toby out of the way and in front of the television set at the top of the stairs so she can have a little peace. But Toby lets Carolyn know he wants to be with her and whomever else is preparing dinner: 'Want to go to the kitchen.' Sometimes he and Carolyn are at odds. Whenever he makes a mess, she tells him: 'Now see what you've done.' Once when Carolyn dropped a glass, Toby was quick to shout out: 'Now see what you've done.' He has a sense of humour and of irony. On one occasion, Carolyn was rushing to leave for the office and Toby did something that needed attention. In haste and exasperation, she walked over to him, her high heels clicking on the tile as she did. Equally miffed, Toby suddenly made the same rapid click-click-clicking sound – 'Same to you', in effect.

Toby often teased Bobby, the Restaks' dog, by saying: 'Bobby. Want to go out, Bobby? … Want to go for a walk? Where's Cooper, Bobby? Want to go play with Cooper?' (Cooper was Bobby's friend, the neighbour's dog.) These weren't meaningless, mimicked phrases. Bobby was around when Toby said them, and Toby was obviously directing the comments to Bobby. And after Bobby died, Toby has mentioned his name only once.

The night at Restak's National Zoo/Smithsonian lecture, Toby's sounds weren't just meaningless 'parrot noise' – which some people who don't realise the intelligence and meaning behind the communication of many animals might suggest. Toby's words and phrases were part of a logical context. He was showing off to the audience. He was attempting to get Richard's attention. Sometimes he was competing with the other bird on stage. If our child does this, we can see the logic behind his or her behaviour. Rather than random or mimicked utterances, Toby's words had meaning, and showed his ability to learn as well as his social and emotional sensitivities. A recognition and understanding of the context is what humans have often missed in our evaluation of animal intelligence. Richard Restak and his family have known this for their twenty-plus years with Toby, and in responding to him logically and contextually, they have set up a feedback system that encouraged Toby's learning and allowed the humans to increasingly understand his communication. As humans come more directly into the interactive loop with other animals, all are affected. We can better appreciate their capabilities as they can bring us closer to their worlds. Learning is a two-way street.

While Alex and Toby as well as other parrots in captivity have impressive levels of understanding, parrots that live in the wild have been shown to be equally remarkable. Many of them live long lives – macaws can live up to 50 years or more. To live so long necessitates a number of survival strategies. These birds need a good memory to remember nesting places, different food sources, mates, and how to best cope with droughts or floods which they might experience several times over their lifespan. Almost all parrots live in flocks that have complex structures and require a high degree of social intelligence. They have to learn a great deal about each flock member and how best to deal with that particular bird. Mated birds often stay together for quite a while, even after the young are raised. In some species, the couple may create a certain song that they sing as a duet, each filling in special notes.

Parrots often use call duetting and call dialects, and the dialects can change depending on the social grouping. For example, parrots may use one dialect if they are interacting with a flock, compared to another they would use around their night roost companions, or a dialect especially reserved for their mate. There is also large-scale geographic variation. It's even thought that some species of parrots can alter their call structure from moment to moment depending on which individual the bird is communicating with. So humans are not unique after all in many of our skills and abilities, and certainly not in what we feel and experience emotionally.

Several parrot species are known to use tools. Male palm cockatoos carve out 'drumsticks' with their teeth and use the sticks to drum loudly on a hollow tree as a way of publicising their territorial lines. Other types of cockatoos hurl small stones, twigs, or other objects down at predators to scare them away. And of course as we've seen in Alex and Toby, parrots have a wonderful sense of humour and play. Australian cockatoos hang on and ride the whirling blades of windmills. Kea fledglings play in groups with sticks, teasing each other and, at times, seeming to simulate a parrot-style sword fight. And they can be a noisy bunch.

In the case of animals in the wild, we don't necessarily know the reason behind some unusual sounds and behaviour. Perhaps sometimes animals just like to learn and make new sounds; maybe they have more in mind. In Karen's neighbourhood, a bird (probably a mockingbird) will imitate the sound of a home burglar alarm or a car alarm – to which humans often react. Sometimes a police car will show up.

Several years ago, as Karen was standing on a small traffic island near the National Gallery of Art in Washington, DC, she suddenly heard a mobile phone ringing. She checked hers immediately, as humans tend to do, even though the ring was very different than her phone would produce. The piercing ring occurred several more times. She was mystified – there in the middle of traffic whizzing by in both directions, and no humans standing around anywhere she could see. The ringing continued – loud, and coming from somewhere nearby. Finally, she looked up into the lone tree on the traffic island. There was a bird looking down at her. It blasted forth in a long mobile-phone-like ring, then looked down; blasted forth again, and then looked down, cocking its head. It was a great mimic job, to say the least. There was no way to know if there was anything else going on; was the bird teasing, as Toby teased the Restaks' dog?

Despite the increasing amount of experimental and anecdotal evidence to show that animals lead lives of an intelligence and complexity hitherto overlooked, at this stage of our scientific research we cannot omit other psychological factors that might be involved as well. Harvard psychologist B.F. Skinner showed in his work with pigeons that 'operant conditioning' can take place when an animal does something and then gets some kind of reward (food, attention) for that particular behaviour. Thus rewarded, it will behave in the same manner again. Operant conditioning occurs constantly in our lives. If a toddler throws a tantrum, for example, and gets a lot of attention for doing so, he or she is sure to throw that tantrum again. (That's why psychologists say it's important to reward children for positive rather than destructive behaviour.) Skinner taught pigeons to play ping-pong through operant conditioning. So there's a chance, for example, that rather than demonstrating sophisticated conscious trickery, the mobile-phone-ringing bird just happened to mimic a mobile phone – then liked the sound (self-reinforcement) or enjoyed seeing humans respond in a bewildered fashion.

However, while we must be wary of reading too much human meaning into animals' acts, especially those which can also be explained in simpler terms, it's undeniable that many species consistently modify their behaviours to enable communication outside of their natural field. There are reported cases in Switzerland of an African elephant changing his lower-frequency African 'language' to that of the female Asian elephants in the same zoo that make a chirp-like call, and in Kenya of elephants mimicking the sound of trucks. One example of truck-mimicking, reported in 2005, was Mlaika, a ten-year-old adolescent female African elephant in a wildlife preserve. By day she speaks 'elephant'; at night she sounds like a truck. In an interview, Stephanie Watwood of Woods Hole Oceanographic Institute said: 'At night, when she [Mlaika] is by herself, she can hear the trucks from a nearby highway and imitates their sounds.' Watwood is the co-author of a study on elephants' copycat behaviour and suggests that elephants have learned to produce sounds outside of their normal repertoire in order to better relate with their neighbours.

Monkey see, monkey do

Some animals mimic in order to better relate with others, other animals mimic to deceive. But all of us have to imitate in some way in order to learn specific behaviour. Animals have what are called 'mirror neurons' – brain cells that have been shown to focus in on and 'learn' behaviour of other animals. Children use these 'mirror neurons' to learn from adults about the world and how to function in it. In Karen's own household, Viva, a six-week-old Maremma (Italian mountain dog) puppy, intently watched the behaviour of the family's fourteen-year-old Belgian/German Shepherd, Schaum. Through this observation, Viva was toilet trained in just two days by imitating Schaum. She also imitated Schaum's method of lying down; the older dog would ungracefully 'plop' down on the floor because of arthritis in her back legs. Even as a healthy adult dog, Viva persisted with this unconventional technique.

In Japan, researchers documented the rapid change in behaviour of an entire troop of Japanese macaques in Koshima, just because one monkey came up with some new ways of doing things. Monkeys saw and monkeys really did. When Imo, a female in the troop, was offered sweet potatoes she started dipping them into the river, presumably to wash off the sand, instead of knocking it off with her hands. Her family copied what she did. Then the whole troop starting dipping their sweet potatoes into the water. Next, Imo would take a bite of the potato and dunk it into the sea, perhaps because she liked the salty taste. Bite, dunk, bite, dunk. The troop followed suit. Every monkey around began biting its own sweet potato, dunking it, biting it, then dunking the potato again. Later, Imo started washing bits of wheat. She'd take a handful of wheat and float it in the water – again, probably to wash off the sand. Suddenly, for the first time in their lives (and perhaps in the history of the entire species) the rest of the troop of macaques became wheat-washers.

Orang-utans are no less observant and ready to learn. Although they are solitary animals, orang-utans are known to be especially adept in watching and learning from other orang-utans and from humans. One orang-utan, involved in a number of laboratory experiments, showed that he could imitate 90 per cent of the body movements that were performed in front of him. These primates are also known to use tools: they find, or fashion, the right-shaped stick to poke insects out of tree trunks or limbs, or to chisel open a termite nest; they use leaves to wipe their faces while eating messy food, as one might use a napkin; they also use leaves as gloves to handle prickly fruit or thorny twigs; they even make drinking cups out of pitcher plants. In captivity, these orang-utans have been known to have language to denote certain objects. And the orang-utans in one primate rehabilitation centre wash clothes with soap and water as they have seen humans do. One young female learned to saw wood and hammer nails.

Karl Pribram, who spent half a century studying and training monkeys and apes, says that one of the most interesting phenomena he saw over the years had to do with the nesting behaviour of pregnant chimpanzees – a behaviour that could either be due to early learning during a critical period or to rapid genetic change based on new environmental requirements. A group of chimpanzees he studied at Yale University had been brought from Africa when they were two months old. Over ten years later, as each female chimp went through pregnancy, she exhibited strong nesting behaviour – gathering and stashing newspapers and anything else she could put in her 'nest'. Chimps who were born in captivity at the Yerkes Laboratory of Primate Behavior never did this. Pribram was intrigued that, even after ten years in captivity, the chimps born in the wild prepared the same type of 'nest' for their baby as their mother had done for them. How did they remember this for ten years? And how could this behaviour be eliminated in just one generation – since no chimps born in captivity carried on this legacy?

From his extensive research work at the Yerkes Laboratory, Yale and Stanford, Pribram tells many stories about monkeys and apes who, like human primates, have their own personalities and moods and compete for recognition and special attention. These non-human primates like to laugh, and have a funny, almost-like-a-human laugh. They don't sing, but they like to dance, especially to a drumbeat. And lab animals take their performance seriously. According to Pribram, 'when male animals are doing well, they get erections. When they're not doing well, they droop. That's consistent with the statement made by a prominent politician that "succeeding is the best aphrodisiac" ... It shows that animals are not like machines doing a routine thing, they're right there with you.'*

In general, these non-human primates enjoyed learning. But if the task became too tedious or was too difficult, they might act up. Frustration could also be a factor. At the Yerkes Laboratory, there was Alpha, the first chimpanzee to have been imported to the United States.

Alpha was a seemingly calm chimp who carried out some very difficult tasks; when things didn't go well, she patiently kept trying, performing a task over and over again until she got it right. Pribram used her in a series of tasks meant to frustrate chimpanzees. He would reward the animals until they performed perfectly. Then he would withhold all rewards for several trials, followed by trials where the reward was reinstated. He timed how long it took for the chimpanzees to resume normal responding. Alpha didn't seem to mind the non-rewarded interspersed trials, and simply continued testing as if nothing had happened. This went on for several weeks until one day a load of faeces came flying over the testing apparatus onto Pribram's head. So much for timing the duration of frustration during individual sessions.

Sometimes the non-human primates would 'go on strike' in the lab: they'd refuse to do anything. On more than one occasion, a divisive chimp instigated a mass rebellion

---

* Clinicians aren't surprised to hear of this monkey reaction. We often have male patients come in who are having erectile difficulties because they don't feel very good about how they are 'performing' in work or in their personal lives. Many human sexual problems are based on a lack of self-confidence.

of chimp subjects against the experimenters. But usually Pribram developed a good rapport with his primates, and often a primate would accompany him, walking hand-in-hand down the hall.

In addition to going on strike, research monkeys have also demonstrated a capacity for deliberate trickery. Other primate researchers have told how one monkey might pretend to be another. For example, one monkey would knowingly get in the seat that another monkey always used and attempt to do what was required of the experiment to get a reward – even making the individualised sounds that the other monkey usually made. The 'imposter' behaviour might occur for a variety of reasons: either to get out of completing another task, to obtain an undeserved reward, to get special attention, as a reprieve from a tedious routine, or perhaps to surprise or irritate an experimenter.

In a number of situations the monkeys would outsmart Pribram. For example, in an operant conditioning experiment, Pribram would carefully record the timing. Eventually, he noticed that something wasn't quite right. It turned out that one monkey wouldn't wait for the reward to drop down from a box at the set interval of two minutes. Instead, the monkey put his hand up inside the hole of the box and grabbed the reward – but just a few seconds before it would have come down, so Pribram wouldn't be alerted.

Washoe and Koko

Washoe the chimp and Koko the gorilla are two of the best known primates in the neuroscience literature. Both have been trailblazers in the study of primate language-learning ability. Pribram advised on some of the research protocols used with Washoe, and was instrumental in bringing Koko to Stanford University and setting up her decades-long learning programme with scientist Francine 'Penny' Patterson.

In 1967, cognitive researchers Allen and Beatrix Gardener set up a first-of-its-kind project to teach a chimpanzee named Washoe American Sign Language. By the time of her death in 2007, aged 42, Washoe could reliably use over 250 signs. Pribram consulted for a number of years on this project and got to know Washoe quite well. In the 1980s, Pribram had a traumatic encounter with Washoe, of which he is reminded daily. In its unfortunate unfolding, the incident showed that Washoe did indeed understand the meaning behind a complex concept, as she signed it clearly in her emotional state. Washoe was caring for a baby chimpanzee at the time of this particular visit from Pribram, and was very protective and a bit edgy. As Pribram was feeding the infant, he ran out of food. He held on to the cage with one hand to reassure Washoe that he wasn't leaving her as he turned and reached over with the other hand to get some food. Washoe smashed his hand against the wiring of the cage with such force that his middle finger was cut. As Pribram was trying to control the blood spurting from his injured finger, Washoe kept signing frantically: 'Sorry. Sorry. Sorry.' Sadly, Pribram lost half of his finger that day.

Koko the 'talking' gorilla was born on 4 July 1971 in the San Francisco Zoo, and named from the Japanese word *Hanabi-ko*, meaning 'fireworks child'. At present, Koko is able to understand more than 1,000 American Sign Language signs and more than 2,000 words of spoken

English. Her caretaking nature was aptly described in Penny Patterson's 1987 book, *Koko's Kitten*; besides All Ball, the kitten featured in the book, Koko has cared for several cats over the years. As a demonstration of Koko's advanced linguistic understanding and emotional reasoning, one spring afternoon at Stanford, as Pribram, Patterson, Pribram's daughter Joanie, and Koko were strolling the campus, Pribram remarked on the beautiful flowers all around and how good they smelled. Penny wanted to show off Koko's language ability and asked Koko to sign to Joanie for her to smell the flowers. Koko understood, grabbed Joanie by her ponytail and pushed her face into the flowerbed. After all, why use language when action works so well?*

Smarter than we thought

How many of us have tried to outsmart the squirrels who steal the birdseed we set out? Karen once devised what she considered a brilliant contraption with a garbage can lid that would go crashing down when touched even slightly as a squirrel attempted a climb to the sunflower seeds twenty feet above. But, no surprise to anyone in this futile struggle, this strategy was foiled! There are whole industries based on outsmarting those animals who want to steal what we have or what we plant.

Psychologists used to insist smugly that only *Homo sapiens* could truly problem-solve, that is, envision future situations and build tools to solve an impending problem.

* 'The answer, by the way,' according to Karl Pribram, 'was given by Freud: talk is an excellent substitute for action and less traumatic.'

But then researchers documented a 'lowly' crow finding and twisting a small piece of wire in order to hook and pull a morsel of food towards him. And carrion crows in Japan use cars to crack their nuts for them. Astounding as this may seem, when cars stop at red lights, crows will put nuts near the car tyres so when they start up again they roll over and crack the nuts. And a number of crow and other bird species are infamous for their creative execution of ways to distract other animals, like cats, so they can steal their food. The green heron uses various kinds of bait that it holds just above the water to lure fish to the surface. And the western scrub jay has been shown to plan by setting a certain amount of food aside based on how hungry it might be the next day. Some people who regularly feed birds claim that birds, like cardinals for example, ask for some food just before nightfall. But they don't eat the food then. They have it at dawn – when they wake up hungry and the humans are still asleep. It seems as if the birds want their breakfast set up early for them.

Now that we've been looking, we're recognising many other examples of planning and even tool-making among a variety of species. We've found that octopuses can distinguish shapes and patterns as well as solve simple problems. In laboratory experiments, common octopuses were able to open jars to get the food inside. And of course, most animals have to plan ways to get their food. Chimpanzees have been documented using numerous tools to get their food. They use stones as nutcrackers and sticks to reach bananas and other fruit beyond their reach. And when they dig, it has been found that, unlike most humans, they are ambidextrous.

And it's not just food. Tools can be used for defence as well as for aquiring food. One chimp was observed carving a sharp stick as a defensive weapon. Another carved a stick to spear a little animal hiding in the hollow of a tree. Recently, a chimp was seen by zoo officials in Sweden planning ahead to throw rocks at annoying visitors. According to a report in the journal *Current Biology*, Santino, a 31-year-old alpha male, started collecting rocks and knocking out discs from boulders inside his enclosed area early in the morning before the zoo opened. Around midday he started hurling these rocks in the direction of onlookers. According to Mathias Osvath of Lund University, 'These observations convincingly show that our fellow apes do consider the future in a very complex way. It implies that they have a highly developed consciousness, including lifelike mental simulations of potential events.' From this highly developed consciousness, it's conceivable that for some animals it may also involve an actual self-consciousness – a true sense of 'self'.

A sense of 'self'

A perennial question with which scientists have grappled is to what extent any non-human animals have a sense of self, whether they self-reflect, and if 'consciousness' is a part of their existence – especially self-consciousness. For many years, psychologists have put animals through the 'mirror test' in which a spot or X is put on the animal's face and a large mirror is placed in front of it. They then watch to see if the animal tries to remove the mark or in any way indicates that it's looking at itself in the mirror. Over the years, chimps and macaques passed the mirror test. Gorillas didn't. Nor did dogs or cats. (But a caveat: it was assumed that an X or 'dirty spot' on a dog or cat's face would bother them, but perhaps they really don't mind a smudge of dirt or a bright red X and see no need to rub it off.) Porpoises passed the mirror test; they would poke at the mirror as if they were trying to remove the mark. Recently, some elephants passed the test and acted in other ways to convince psychologists that they do engage in self-reflection and possess a sense of self. Then, recently, a small-brained bird passed the test. Now scientists are confused. This isn't going as expected. Small-brained songbirds aren't 'supposed' to have a sense of self. The mirror test is the best we can think of thus far, but maybe we're asking the wrong question – or setting up inadequate experiments. Maybe many of our old assumptions about animals are wrong. Time and more testing will tell.

The eyes have it

Another area that's being given attention is that of eye contact among the same species and also between different species and humans. Although dogs aren't supposed to be very good at eye contact – sometimes we're warned that a direct stare at a dog could be perceived as an act of aggression – many of those who have dogs speak of the prevalence of direct eye contact they have with their pets and the extensive amount of information that's shared. Is this an outcome of dog–human interactions over the years? Wildlife enthusiasts speak of the frequent eye contact they have with a number of animals around, such as deer and squirrels. Karl

Pribram argues against those who assert that monkeys and apes are 'clueless' when it comes to understanding what eyes communicate. The only cluelessness seems to be that of the humans who haven't yet figured out the communication system. As this topic is studied in more depth, it's probable that scientists will find that there's a great deal of information being exchanged from eye contact, particularly among social animals – humans included. Sometimes that eye contact has to be more subtle, such as in a number of human cultures, and sometimes it's more direct.

In a recent publication in *Current Biology*, Cambridge University researcher Nathan Emery showed that the jackdaw, a close relative to crows and ravens, gleans information from eye contact and through that information figures out what's on a person's mind. Emery found that when a stranger looked at a particular portion of food, the jackdaw took quite a long time to go after it, possibly because the bird inferred that the person was interested in the food. But if a person familiar to the jackdaw looked at the food, the bird inferred that the person was encouraging it to eat the food and went after it immediately. Jackdaws were also able to follow a person's finger pointing to food, and go to eat it. Finally, in an experimental set-up in which a person stood in front of the bird and looked first at the bird and then at the food, the bird used that cue as well to find the food.

This study alludes to people who are familiar to the jackdaws, but it doesn't explain what cues the birds use to evaluate this familiarity. Could it be the people's faces? University of Florida researcher Douglas Levey concludes from his study of mockingbirds that they are indeed able to distinguish faces of humans after only 60 seconds of exposure time. He had a number of students approach the nests of mockingbirds and gently touch the edge of the nests. The students returned the next day to do the same, and the next day and the next day. The third and fourth times the students approached the birds' nests, the mockingbirds attacked – rushing out screeching at the students and even hitting them on the head. On the fifth day, a totally new group of students was sent to approach the birds, and the birds barely reacted to them – demonstrating that the mockingbirds could tell that these weren't the same faces. Humans might not be able to tell the difference between some birds of the same species, but at least the mockingbirds seem to know us apart.

Memory and counting

As most of us grapple with trying to remember faces and so many other things in our lives, it doesn't necessarily make us feel better to know that a recent study in Japan showed in a specific test of memory for placement of numbers that chimpanzees did better than college students. The five-year-old chimps had already been taught their numbers and how to count from 1 to 9. Then they were shown the numbers in certain patterns on a screen and tested on remembering the particular sequence shown. The young chimps did better in speed and accuracy on the memory tasks than college students did.

But one bit of information might help us all put our memory concerns in perspective. Animals also get confused and forget where they put things. This is illustrated in the finding that one poor bird – forgetting in which of the three holes in a flowerpot it was making a nest – ended up constructing the beginnings of nests in each of the three holes. We know that our dogs forget where they leave the ball or hide bones sometimes, and probably even the smartest squirrel is remiss at times in finding that hidden nut. So let's not be too hard on ourselves; we lead busy lives, and are especially prone to forget information when we're on 'overload'.

More and more research is being done on animals and counting. It would make sense that animals know how to count. They have to keep track of their young, remember specific landmarks – sometimes from the air – and deal with herds and flocks, and forage for food. Some researchers believe that when they forage for food, ants have to remember how many steps they are away from home. Research is often very painstaking, but imagine capturing an ant and putting stilts on its tiny legs to document how it expects to walk a certain amount of steps to return home, and what happens if its little legs are lengthened to change the length of each step. After attaching pig bristles to each of the ant legs to make them longer, scientists in Ulm, Germany and Zurich calculated that the ants went past their homes by the precise distance expected if they took the same amount of steps returning home as they had taken going from their home to the food source in the first place. The researchers concluded by this and other manipulations of leg length that ants do indeed count their steps.

Not all scientists agree that ants actually count their steps, because such an idea is way beyond the numerical competence found in human babies and other animals. Although it's very difficult to study numerical skills among animals, many researchers have spent a professional lifetime doing so. The representation of numbers has been especially important in psychological studies because it connotes relatively high-level thinking. Unlike describing one entity, numbers are abstract. They are a construct that describes a set of entities that can be objects, like how many cups are in a row; they may be events, like how many times a cat rolls over or how many times one gets rewarded for the same behaviour; or they may be non-physical entities.

Animals can be trained in a certain type of use of numbers. For example, a rat can learn to press a lever a certain number of times, like 45, to get a reward. And bees recently have been shown to be able to substitise – estimate small quantities. Jürgen Tautz and colleagues of the University of Würzburg in Germany trained several groups of bees to fly through a Y-shaped maze to get a sugary reward. After being shown a two-dot sample picture, the bees had to choose whether to go to the portion of the maze with a two-dot pattern or the side with a three-dot pattern. After a certain amount of repetition, the bees were able to choose the correct, two-dot pattern 80 per cent of the time. When the researchers made the patterns more complex, bees still were able to tell the difference between two and three and three and four. Beyond four, the bees didn't seem to be able to make distinctions. As with non-human primates, four seems to be a threshold. Tautz finds the division between 'up to four' and 'a lot' intriguing, since he says that it's mirrored in many human cultures. And Lars Chittka of Queen Mary, University of London, says the study shows that even simple animals can extract quantitative features from visual signals.

A number of other studies have shown that both humans and animals have the ability to mentally represent numbers and compare them – for example, they can discriminate between four objects and eight objects. And for human babies and for other animals including salamanders, the threshold for precisely remembering a certain amount of numbers is four or five. One of the first studies to show that animals could also perform mental arithmetic was done in 2007 by Elizabeth Brannon and Jessica Canton of the Duke Center for Cognitive Neuroscience in Raleigh, North Carolina. They placed a macaque monkey in front of a computer screen that displayed a variable number of dots; then there was a second screen showing another number of dots. For example, the first box on the screen might contain two dots and the second box might contain five dots. Then a third screen appeared containing two boxes. One box contained the sum of the first two sets of dots (seven) and one contained a different number of dots (say, eleven). The monkeys were supposed to touch the box displaying the correct sum of the first two sets. College students who took the same test, and were asked to choose the correct sum without counting the individual dots, were correct 94 per cent of the time. The monkeys were correct 70 per cent of the time. Most subjects responded within a second and found the task much more difficult if the sum number of dots and the other picture of dots were close in number and were of higher numbers, such as eleven dots and twelve dots.

★

So now we know that animals can discriminate between numbers, can count, and can do mental arithmetic. They can make tools and plan ahead, and there is also research showing that monkeys even rationalise their behaviour. We're still trying to figure out which animals have a sense of self and their level of consciousness. Most of all, we're realising that animal traits and intelligence are not simply hierarchical, increasing in complexity from one-celled animals to humans. Birds, for example, are much more capable of 'thoughtful' behaviour and are more intelligent than scientists had supposed. Research shows that environmental demands as well as interaction and learning among different species create a spiral of constant change and increased complexity at every level. It will be very interesting to learn more about the intellectual capacity of the newly studied 'brainy echidna'.

9

Eavesdropping and Deception

Oh what a tangled web we weave,
When first we practise to deceive!
Sir Walter Scott, *Marmion* (1808)

Among wildlife there are many documented cases of eavesdropping and outright deception. Animals listen in or trick each other to survive – to get a mate, to keep a mate, to protect the family, to acquire and keep cherished food. They also use deception to get back at others and just plain to get attention. Roosters, for example, will lie to get attention by pretending that food is nearby to attract hens to their part of the yard. Frogs change their croaks or even stutter to get what they want; opossums play dead to avoid danger; monkeys may pretend to be somebody else; and elephants mimic the sound of trucks.

Eavesdropping

Listening in has always been fun to do, and often serves a serious purpose as well. Many animals know the signals of other species or the special dialects and 'secret' signs of those within their own group. For example, in the midst of group bird chatter, a bird from another group – or a predator from a different species – might be listening in. In fact, many animal calls and conversations are visited by outsiders. Just as is the case with humans, information is power, and a bit of gossip won't hurt.

Bats are known to eavesdrop on their neighbours as well. An echolocating animal gives away a lot of information about itself – such as its position and flight path – by the timing and structure of its echolocation pulses. High-intensity echolocation pulses may reveal the identity of the bat giving off these pulses and also the presence of a food source – valuable information for a sneaky listener. Or the eavesdropper may not zero in on only one bat, but listen for a high density of echolocation pulses that might indicate a gathering around a food source – a lot of noise means a crowd of bats and a high-quality food source. The bat *Myotis lucifugus* has been found to do this. Listening for the signals of others that are foraging might work in an opposite manner as well. Instead of joining a group of foraging bats – the big crowd at the 'bat restaurant' – some bats may decide to space out a bit to look for another, less crowded food source. For example, in some larger bat species, adults use echolocation pulses produced by bats that are foraging as an indication for the listeners to move away at least 50 metres from that particular feeding ground. This type of transfer or gathering of information from echolocation sounds can be considered as an indirect one-way form of communication.

Eavesdropping for a bat is a complicated activity – especially among bats that echolocate. They have to focus enough on their own signals to navigate correctly, yet may pick up on the vocalisations of fellow bats that are emitting echolocation pulses at a slightly different frequency. This is intriguing, because several neural mechanisms have been revealed that show how bats can tune in to their own frequency to prevent their echolocation signal from being jammed by the cacophony of thousands of other bats also emitting many high-intensity pulses in close proximity and at the same time. Yet they manage to do this effortlessly. One possibility is that they can process the two types of signals in parallel by the two opposite halves of their brains. When one – the right – or even both sides are busy processing echolocation signals, the left side of the cerebrum may be easily alerted to communicative signals or echolocation signals of others that have communicative value. Studies to test this hypothesis are currently under way in Jag's laboratory. Earlier experiments have already demonstrated the specialisation of the left half to process communicative sounds, just as the left side of the cerebral hemisphere in humans is specialised to process speech. These types of advancements in non-human animals provide a window into the origins of many of our own amazing abilities.

Even frogs and toads engage in social eavesdropping. It has been found that females eavesdrop on male–male interactions and use the information gathered to make mating decisions. In both birds and frogs, the advertisement call is directed at both males and females – it's not just a mating call, but a territorial and competitive call for information. By following the interaction between two males, female frogs are able to extract information about who is the leader. This doesn't mean, however, that the leader is necessarily the first choice as a mate, since many females prefer followers. It has also been found that females may prefer males that jam the call of other males. The fact that these types of decisions can be made by the lowly frog suggests several possible conclusions. One is that these decisions are important in sexual selection. And the second conclusion is that higher-order brain centres and thinking usually aren't necessary to make such decisions; the neural circuitry for making decisions was perfected early on in evolution and carefully embedded into the brainstem and midbrain. And a totally different conclusion is that thinking and decision-making for some animals may not be limited to the type of brain circuitry or processes on which the majority of neuroscientists are focusing. Scientists are starting to explore other scales and levels of information and processes, in addition to physical circuitry.

Sex, lies and videotapes: the audience effect

How we talk to a friend and colleague and how we talk in front of an audience, or even among a group of three or four persons, is quite different. Not necessarily in what we say, but how we say it. The modulating effect of an audience is equally pronounced among communicative interactions in animals. Recent research indicates that animal vocalisations can refer to objects in the outside world; and the sender, depending on the type of audience involved, may control call production. Wild male Thomas's langur monkeys (*Presbytis thomasi*), for example, call when confronted with a model of a tiger only when in a group, and not when they are solitary. This was the first experimental study on wild primates to demonstrate that the presence or absence of an audience influences calling behaviour. The results indicated that males in mixed-sex groups give more loud calls than solitary males when exposed to a predator model.

And of course, when there's a group, there are always individual interests that promote trickery. Just as some animals indulge in eavesdropping as an indirect assessment for foraging or mate choice, others use deception – a more active mechanism for achieving the same general goals.

It was at an American–German student exchange workshop that Jag had his first demonstration of deceptive calling in birds. Professor Mazakazu (Mark) Konishi had just finished his keynote talk in the morning, describing how the studies of Peter Marler and colleagues show that roosters can produce deceptive calls. Thus, a bantam rooster with food is more likely to vocalise if there is both food and a hen nearby, but less likely to call if there is food and another male nearby. This type of audience effect goes further. A hen may even prime a rooster to call and peck at a twig even in the absence of food. The workshop was held at a small farm in Darmstadt, Germany, and the talk was followed by a tea break when all the scientists got the opportunity to sit outside and watch the pigs and hens trot around. Suddenly Jag heard the cluck, cluck, cluck of a rooster and looked up to see a large rooster standing next to a tree trunk and pecking at the ground for no reason. Immediately, all of the hens foraging over an area of a few square metres stopped what they were doing and ran towards him. The rooster's trick had worked. With his head held high and his chest thrust out, he took a few royal steps forward. He had succeeded in drawing their attention and attracted them all towards him. The hens had mistaken the rooster's communication calls as a sign of high-quality food on the ground, perhaps corn or grain that they could all peck on – just as the rooster had intended.

Several species of primates produce food-associated vocalisations upon finding or consuming food. Brown capuchin monkeys give distinct calls when they find food. Studies on other species had shown that both the amount of food and the nature of an audience can influence the type of call that goes out. Therefore scientists predicted that capuchins would adjust their food calling for those variables as well. They tested twelve female capuchins in two food-quantity conditions (large and small) and four audience conditions with a control (higher-ranking female, lower-ranking female, high-ranking male, entire group, and alone). All subjects called more for larger amounts of food in contrast to smaller amounts, and the highest-ranking females called less than others. Subjects called more in the presence of a group than for any other audience, and this applied most strikingly to high-ranking subjects (kin) rather than group size. Therefore, food calls by brown capuchins not only reflect a simple response to the food, but are influenced by multiple audience effects as well.

Tufted capuchin monkeys are cleverer still. They produce two food-associated vocalisations (the 'grgr' and the food-associated whistle series). One investigator placed new food sources (half-pieces of banana on pre-installed feeding platforms) and explored the factors that affect the production of food-associated calls in a wild group of tufted capuchins. Finders of these platforms called in 81 per cent of the discoveries when the platform contained fruit, but in 0 per cent of cases when the platform was empty. Males and females of all ages and dominance ranks gave food-associated calls when discovering a platform with fruit. The probability that a finder gave food-associated vocalisations was lower during the period of food scarcity and when the platform contained a small amount of bananas (three pieces as opposed to twenty or more). There was an effect of the audience on the delay (or 'latency') to give food-associated calls. The time elapsed until the finder gave the first food-associated call decreased with the presence and density of nearby individuals and increased with the distance from other individuals to the platform. The latency to call was longer for females than for males. The audience effect and the effect of the sex of the finder are consistent with the hypothesis that capuchins use these vocalisations deceptively by withholding information about the presence of a food source. The scientists concluded that by increasing the latency to call, finders of new food sources can obtain a larger amount of food and thus reduce the costs associated with calling.

Deception in its many forms

Aesop's *Fables* captured the popularity of deception in the animal world. Throughout history many storytellers have done the same. The metaphors of animals outsmarting each other touch the core of our experiences, making detectives of us all, as well as ensuring that we become shrewder and more cautious. Sometimes we laugh, as in the story from Joel Chandler Harris's *Uncle Remus* folk tales of the captive Brer Rabbit who begged Brer Fox not to throw him into the briar patch. And of course when the fox threw the pleading rabbit into the thorny patch anyway, the rabbit danced and jumped around with satisfied glee. Rabbits grow up in briar patches; it's their safe retreat. The rabbit had finally 'outfoxed' the fox.

Sometimes deception is used to stay alive, as with bright colours or camouflage, or enhancement of size, to defeat or scare a predator. Sometimes deception is employed to seduce a mate or to protect the young and the community. Deceptive ploys are also carried out to get rid of rivals – such as the electric fish that jams the frequency signals of a rival male as seen in Chapter 1, or the bullfrog that changes his croak to simulate a female frog's or to seem like a much larger male.

Researchers caught the male Atlantic molly fish in a surprising form of rival deception. Martin Plath of the University of Potsdam and the University of Oklahoma and his colleagues set out to find whether a fish's social environment affected mating preferences. They discovered that the mere presence of a rival had a robust effect. In the experiment, a male initially would choose larger females over smaller ones, and females of its own species instead of asexual females of the Amazon molly, a related species. When a second male was put into the tank, however, the first male abandoned his initial choice and went to a second and not so 'appealing' female. Why? Well, it turns out that males are known to copy the mating choices of another male. So this appears to be a case of the first male pretending to choose a prospective mate, fully expecting the rival male to try to steal her away. The experimenters found that this switching happened almost every time the new male was introduced to the tank. It just couldn't have been that the preferred female would become less attractive in every case. To complicate events further, the first male even became sexually active with the non-preferred female. The misleading ploy by the first male could have two advantages. It may reduce sperm competition with other males, as surrounding males may use this public information and follow the original male's mating choice. It may also in an interesting way perpetuate the Amazon molly species. Although Amazon mollies are asexual, they need insemination by the male Atlantic mollies to initiate the process of embryogenesis, which results in female-only offspring.

At other times, it appears that plain old laziness may be behind deceptive behaviour. Some animals just don't want to put in extra effort, and find a way to have someone else do the work. Sometimes a whole group perfects this ploy. One type of butterfly could be considered the quintessential moocher.

Most of us would like to live like royalty. Scientists recently discovered a species that can move into the 'palace', usurp the queen, and trick the 'servants' into taking care of it. In short, this butterfly lives a youth of luxury. Researchers found that alcon blue butterfly caterpillars infiltrate red ant colonies by mimicking the scent – the larvae chemically 'smell' like ants – and the raspy sounds of the red ant queen. Once inside the colony, the interlopers feast on the delicious food fit for the 'queen'. Sometimes the caterpillars are so convincing that the ants exile or kill the actual red ant queen – erroneously labelling her as the invader.

But the plot can thicken, turning into a double-layered deception. Jeremy Thomas from the Centre for Ecology and Hydrology in Dorset, UK has found that there may be a lurking parasitic wasp (*Ichneumon eumereus*) ready to sneak into the nest. The wasp releases a pheromone of its own that repels the ants and even makes them attack each other. In the midst of the bedlam, the wasp searches for and finds the caterpillar, lays an egg deep inside its body and exits the nest. Then all seems to return to normal, with the ants feeding and nursing the caterpillar as one of their own. But when it's time for the caterpillar to become a chrysalis, it's eaten by the grub wasp within. Instead of a beautiful alcon blue butterfly, a wasp emerges.

As unfair and unpopular as lazy species or individuals may be, some scientists suggest that the laggards are tolerated by larger groups, because sometimes the fat and lazy ones may be called upon in crises. This is the case with one type of mole rat; it turns out that the fatter and lazier rats are able in tough times to dig their own exceptionally large tunnels or break into existing tunnels to rescue other rats that need help, or to mate with females that had not been previously approached.

It's difficult to say whether the European common cuckoos that lay their eggs in the nests of other species would fall into the lazy category or if their deception occurs for other reasons. Not all cuckoos lay 'orphan' eggs in other birds' nests. Many species of cuckoo build their own nests and raise their own broods like most birds do. And those family cuckoos tend to lay the usual sorts of white eggs. But the brood-parasitic cuckoo species – including the European common cuckoos – that deposit their eggs in the nests of other birds can vary the egg colour to match that of the 'foster family' eggs. The renegade eggs also have thicker shells so they don't break during the 'drop off'. The cuckoo chick hatches earlier than the chicks of the host birds and grows very quickly. The intruding baby then usually takes over the nest by throwing out the host eggs or evicting the host chicks. Some potential host birds try to keep unwelcome cuckoo eggs out of their nests by building different types of barriers. Host birds will also form mobs to drive the parasitic cuckoos out of the area.

There are also a number of animal species that break the 'rules'. Marine biologists have long known about the renegades of a number of fish species called 'sneaker fish' or 'sneaking fish'. These are males that play the mating game their own way, sometimes because they are smaller or weaker than other males, perhaps sometimes because they don't want to fight, outwit or directly compete with aggressive rivals. So they 'sneak around'. They find clever ways to get to the females or the pools of eggs and leave their sperm. Often other males unknowingly end up taking care of the progeny of the sneakers. But the females usually know what happened. In fact, some biologists think that female fish might encourage such sneaking around so they can retain control of the courting ritual as well as the fate of the babies – they don't have to put up with bossy, overly aggressive males.

There are many examples of renegade males throughout many species. Until recently, it was thought that the main reason they would sneak around and/or disguise themselves as females was to avoid or outwit competitor males. But a recent discovery about garter snakes has made scientists rethink some of their boiler-plate conclusions. On coming out of hibernation, a number of garter snakes pretend to be females. Instead of moving around on the fringes of the snake group, sparring with other males to eventually win over mates, some males will curl up in balls of snakes with other pretenders and with actual female garter snakes. Scientists now think they do this for the simple reason of staying warm. It's often still quite cold outside, and warmth is an important aspect of staying healthy. And snakes are cold-blooded – they take on the temperature of their surroundings. At the same time, these female-pretenders are close to the real females. So when these cunning individuals are warm enough and ready to find a mate, all they have to do is publicise who they really are. The lesson here for scientists is that animals may have reasons for behaving in certain ways that we haven't yet figured out.

10

Rhythm, Song and Dance

On with the dance! let joy be unconfined;
No sleep till morn, when Youth and Pleasure meet
To chase the glowing hours with flying feet.
Lord Byron, *Childe Harold's Pilgrimage* (1812–18)

Since the invention of functional magnetic resonance imaging, or fMRI, scientists have been studying the brain regions involved in music perception. Although we all appreciate the importance of music in our daily lives, the scientific basis of why we do, and why we even get up and dance to a tune, is not well understood. Only recently, scientists have discovered that specific areas in the human brain show increased blood flow, and are stimulated and active in some manner, when a person hears a beat versus a melody. It had been thought that dancing to the rhythm of sounds was something that only humans did. Each movement appeared to set us apart from the other human cultures and especially from the other animals whose movements we frequently mimic in our dances. We assume that only we have the capacity to enjoy these moves, but some scientists report that other animals, in particular birds such as parrots and even elephants, can dance to the beat of a drum. They might not always get a spot in a musical, but some animals certainly can dance.

Bobbing to the beat

A recent *World News* story reported that there's new scientific evidence that parrots can dance to a tune and even rock and roll. The report was based on a study of the movements of Alex the African grey parrot who we met in the previous chapter, and Snowball, a medium sulphur-crested Eleanora cockatoo (*Cacatua gaterita eleonora*), as they 'danced' to the beat of human music. Adena Schachner, a graduate student in psychology at Harvard University, and her colleagues showed that these birds' movements were more lined up with the musical beat than we'd expect by chance. There was strong evidence that Alex and Snowball were synchronising with the beat when the music was sped up or slowed down – something that had not been documented before in the laboratory in other species.

Some scientists surmise that one reason for the birds' dancing accomplishment may be the ability of vocal mimicry that Alex and Snowball have in common with humans. Perhaps the same brain mechanisms that allow for vocal mimicry also allow an animal to keep a beat. For both mimicry and dance, you're listening to a sound and constantly monitoring both your output (the vocal sounds you make or the tapping of your foot) and the sound coming in through your ears.

Schachner and the other researchers tested this idea by searching the YouTube databases for videos of animals, including mimics and non-mimics (such as most cats and dogs – although canines do mimic sirens), moving along with a musical beat. They tried to rule out potentially 'fake' videos where music was added afterwards or the animal was following some visual movement. All the animals whose speed the researchers judged to match that of the music, and that kept in time with the beat, were vocal mimics – including fourteen parrot species and one elephant. According to Schachner, 'some of the brain mechanisms needed for human dance originally evolved to allow us to imitate sound'.

Dancing and other rhythmic movement would fall under the category of rhythmic behaviour, which, according to Cambridge University researcher John Bispham, 'implies an order of output with reference to a sustained attentional pulse'. Scientists are grappling with various aspects of rhythm. There's one's internal awareness of and internal initiation of rhythmic behaviour; and then there's the awareness of external, environmental rhythms and possible entrainment with those rhythms.

How and in what circumstances animals are able to perceive rhythm is difficult to study. Franck Ramus, Marina Nespor and Jacques Mehler showed in 1999 that, like human babies, cottontop tamarin monkeys were able to tell the difference between two distinct foreign languages based on the rhythm of each language. Recently, researchers found that rats have the same ability. Other species have the ability to not only perceive but also generate a rhythm. Songbirds have rhythm in their song, while others such as woodpeckers, that lack the ability to generate songs, employ a long-distance non-vocal acoustic signal aptly referred to as drumming. Drumming is unusual in that a separate instrument in addition to the bill or an appendage is required to produce the rhythmic signal. Studies conducted only a few years ago have shown that woodpecker drums encode information for species recognition. The fact that so many animals send out rhythms such as song or drumming could be taken as proof that rhythm perception is quite common among a variety of species. Many animals, including palm cockatoos, woodpeckers and kangaroo rats, 'drum' on hollow objects to signal to others. Gorillas have been observed drumming with both hands, as have chimpanzees and bonobos.

Some researchers are also attempting to identify the types of music or rhythms that may be preferred by different species. Josh McDermott of the Massachusetts Institute of Technology and Marc Hauser of Harvard University found that, again like human babies, cottontop tamarins and marmoset monkeys preferred lullabies and slower tempo tunes to fast dance tunes.

Fish seem to respond to Mozart. Sofronios Papoutsoglou of the Agricultural University of Athens found that carp that were exposed to 30 minutes or more of Mozart's *Eine kleine Nachtmusik* seemed to improve in their growth and were found to have reduced levels of brain chemicals (neurotransmitters) associated with stress. Many pet owners notice how their particular pet responds to certain music or song. Sometimes it depends on the circumstances. Snowball the cockatoo's favourite is the Backstreet Boys. And remember that Alex the African grey parrot loved to dance to The

Mamas & The Papas, but when he was upset only Haydn's cello concertos could calm him. Interestingly, although Alex was quite a talker, he didn't sing to 'California Dreamin''.

Songs

Except for the multi-talented manakins, birds generally don't sing and dance at the same time; and among all the different animals, it's mostly birds such as parrots that produce speech sounds like humans. This is despite the fact that their sound-producing organ, the syrinx, is organised differently from our vocal apparatus or sound box. Our vocal cords open and close rapidly within the pharynx, the alimentary canal cavity between our mouth and nasal passages and our larynx and oesophagus. In contrast, birds have a set of delicate membranes within the syrinx that vibrate when air flows over them. Unlike our sound box which is located at the top of our trachea, the bird's syrinx is set much lower down, at the junction of the two bronchi or air tubes leading to their lungs. This means that the syrinx has two potential sound sources, one in each bronchus. The membranes on each bronchus produce separate sounds that are then mixed when fed into the higher vocal tract. This complex design means that birds can produce a far greater variety of sounds than humans can.

Before going further into birds' extraordinary breathing processes and how they allow them to sing so beautifully, there's a historical perspective worth visiting; it shows how birds' breathing capacity has also influenced their ability to fly and navigate to just about every place on earth. There are about 9,700 species of birds spread all over the globe, from terns in the Arctic to penguins in the Antarctic. Modern birds are thought to originate from an extinct group of reptiles called theropods that were common inhabitants of the earth approximately 200 million years ago.* The strong and continuous breathing most likely present in these early ancestors enabled them to fly long distances, take off and land, and manoeuvre in the air, deftly negotiating potentially dangerous wind gusts without crashing. This probably contributed to the extraordinary singing ability of present-day songbirds.

This remarkable history means that birds today are never out of breath, because they have a complex system of air sacs connected to their lungs, and air can flow back and forth in between the spaces. Oxygen transfer to the blood is therefore a continuous process and an efficient way to use up all of the oxygen during inhalation as well as exhalation. Birds literally swallow the air through their oesophagus, which is connected to the air sacs and lungs. A canary may take 30 mini-breaths per second to replenish its air supply. These mini-breaths are synchronised with each syllable or note that the canary sings, enabling it to sing continuously and effortlessly for several minutes. And such mini-breaths and constant breathing give songbirds a decided performance advantage in the opera of life.

Singing is the most generic form of music that was perhaps practised by cavemen long before they discovered the joy of producing tones by blowing air through a hollowed

* Research suggests that the earliest flying reptiles swallowed small pieces of volcanic rock and could breathe out flammable gases like hydrogen produced in their own bodies. Their ingenious 'fire breath' was used as a defence against predatory reptiles.

bone. That inspiration may have come from listening to the birds chirping, whistling and singing to establish their territories at the onset of a new mating season. Fernando Nottebohm at Rockefeller University was the first to discover that every spring, the song control nuclei in the brains of songbirds enlarge, shrinking again in the autumn. The waxing and waning of this well-defined brain structure coincides with the singing activity in birds. Such an important finding changed our thinking forever about how malleable and changeable one's brain can be. Opera singers would be delighted if they could do the same every time they have to perform in front of an audience.

Singing isn't restricted to songbirds. Deep under the ocean, whales sing as well. Their songs carry over hundreds of kilometres of water to their mates or distant relatives swimming within their own group, because sound travels better under water than in the air. All species of whales produce long drawn-out sounds that can be detected thousands of kilometres away. The songs of the humpback whale are among the most complex in the animal kingdom. Their beautiful and varied songs consist of repetitions of a series of sounds with considerable precision. Researchers have now mathematically confirmed that whales have their own syntax that uses sound units to build phrases that can be combined to form songs that last for hours.

Until recently, only humans demonstrated the ability to use such a hierarchical structure of communication. Research published in the *Journal of the Acoustical Society of America* offers a new approach to studying animal communication, although the authors don't claim that humpback whale songs meet the linguistic rigour necessary for a true language. Ryuji Suzuki suggests that 'humpback songs are not like human language, but elements of language are seen in their songs. For six months each year, all male humpback whales in a population sing the same song during mating season. Thought to attract females, the song evolves over time.' Suzuki and co-authors John Buck, an electrical engineer who specialises in signal processing and underwater acoustics at the University of Massachusetts, Dartmouth, and Peter Tyack, a biologist at Woods Hole Oceanographic Institution, also in Massachusetts, applied the tools of information theory to analyse the complex patterns of moans, cries and chirps in the whales' songs for clues to the information being conveyed. Suzuki designed a computer program that enabled scientists to analyse acoustic characteristics and classify the structure of the whales' songs, measuring a song's complexity. The computer-generated model as well as human observers agreed that whale songs are hierarchical, confirming a theory first proposed by biologists Roger Payne and Scott McVay in 1971.

The structure of the humpback whale song is repetitive and rigid. The whales repeat unique phrases made up of short and long segments to craft a song. There are multiple layers, or scales, of repetition, denoted as periodicities. One scale is made up of six units, while a longer one consists of 180–400 units. The combined periodicities give the song its hierarchical structure. Suzuki and his collaborators were also able to show how much information can be conveyed in a whale song. Despite the 'human-like' use of hierarchical syntax to communicate, Suzuki found that whale songs convey less than one bit of information per second. By comparison, humans speaking English generate ten bits of information for each word spoken.

Like songbirds and some species of whales and dolphins, a few species of bats have been shown to sing and engage in courtship displays. Although, as we saw in Chapter 3, the navigational echolocation sounds produced inside water are quite different in their acoustic structure from those produced by bats flying in the air, the communication sounds under the water and in the air are relatively similar. So, if bat sounds are played back from a tape recorder at slow speed, they can sound very much like some bird calls, and when slowed down even more, they don't sound that different from the sounds produced by dolphins and whales. Singing in bats is strongly seasonal in nature and probably establishes the territory of the foraging bat. False vampire bats, *Megaderma lyra*, engage in song flight behaviour. This behaviour is displayed only by the dominant males and can occur at any time of the year. It consists of a stereotyped flight pattern that is continuously accompanied by vocalisation. The behaviour is clearly one aspect of courtship and is interestingly directed only at non-lactating females within the group.

In the only study describing this behaviour, it was also elicited by the introduction of new females to the colony. The song flight in *Megaderma* consists of three stages that have been named the introductory, the advancing and the final flight. Spectrally distinct segments or strophes accompany each of these stages. These bats also interact socially in various contexts and emit different types of communication sounds. Unlike some of the other species studied, many of the auditory communication behaviours observed in *Megaderma* occur in flight. One of these behaviours is labelled 'grumbling flight' and involves several bats hovering for a few seconds in a head-to-head formation emitting a series of short, downward frequency modulated (FM) sounds that may terminate in a shallow downward FM. The spectral structure of this sound sequence is remarkably similar to the checked downward FM (cDFM) sounds observed in moustached bats. This may be preceded by an initial side-to-side flight by two or more bats in an excited state. Reports on social interactions during flight are usually based on chance observations, apart from honk calls which are emitted when flying bats are about to collide.

In the sac-winged bat, *Saccopteryx bilineata*, singing is used to establish roosting territories. This species employs an unusually large vocal repertoire. Males emit tonal calls while interacting with females, and other types of calls consisting of composites when actively defending their territories. Their songs consist of short repeated tones that don't appear to have any obvious context other than advertising the quality of singing by a male. These bats don't hang in large colonies. Rather, each bat has a favourite roost in a tree from where the males sing complex songs to attract the females.

Mice have also been found to sing. Now that scientists have the technology to pick up their high-frequency sounds, some of the sounds turn out to be songs. In 2006, field biologist Martina Kalcounis-Ruppell recorded a singing mouse in Carmel Valley, California. According to Mark Stromberg of the Hastings Natural History Reservation, the recordings of the shrill mice chirps are a big hit in his circle of colleagues. 'This is the first time anybody has recorded mice making sounds like this', said Stromberg – songs that are 'complex' and melodic. The mouse singing sounds like that of a songbird, but is above the frequency range of most people. Stromberg said that the singing mice are of European descent and common in central California, especially in areas of Monterey County such as Carmel, Pebble Beach and Big Sur. We don't know yet if mice sing duets or in a chorus.

The dawn chorus: group dynamics

One of the greatest pleasures many of us experience is to sit in a beautiful garden listening to and observing birds. Growing up, Jag enjoyed this on a daily basis in his back yard in New Delhi, India, where birds abound. He remembers vividly how, early in the morning at the beginning of each sultry summer day, hordes of shrieking parrots would come to bite into the ripening papayas. The parrots' loud screams, together with the cawing of the crows, would drown out the twittering of the myna birds and the chirping of the house sparrows. The morning din of the birds gradually died down as the sun rose above the horizon and the birds sought the shade of the shrubs and the tall trees. Only later in college when taking a course on animal behaviour did Jag realise that there was a hidden, but precise, pattern and motive to all of this.

Group dynamics is a phenomenon in psychology that has been the focus of a number of sociological and communication studies. A group is composed of two or more individuals who are connected to each other by social relationships. Because they interact and influence each other, groups develop a number of dynamic processes that separate them from a random collection of individuals. For centuries, scholars have been fascinated by groups – by the ways they form, change over time, dissipate unexpectedly, achieve great goals, and sometimes commit great wrongs. The tendency to join with others in groups is not unique to humans. Psychologists, zoologists, neuroscientists, mathematicians and other academics are studying the grouping of animals such as their flocking, swarming, and shoaling behaviour, as well as their groupings for foraging and mating. Researchers are looking at the effects of circadian rhythms and responses to light and natural surroundings on group behaviour.

The dawn chorus is one of the marvels of nature. Birds all over the world – from English woodlands to tropical rainforests – show the greatest amount of singing activity around dawn. Scientists are just beginning to study this daily eruption of activity and song. According to ongoing studies by John Burt and Sandra Vehrencamp at Cornell University, the dawn chorus is not just an important ritual, but a time to re-establish territorial boundary patterns and nest-site movements. The chorus represents a multi-directional interactive communication network between individuals of a species and probably across different species. Vehrencamp and her colleagues studied the dawn chorus using a distributed array of microphones placed in the field. They propose three basic network components within this communication framework. The first possibility is that of a broadcast network, in which at least one sender produces undirected one-way signals that are received by potentially many receivers. Another possibility is that the chatter represents an eavesdropping network, in which two signallers interact and eavesdroppers obtain relative information about each one interacting. This secret listening in may be considered a passive form of communication or gathering of information from the vocal signals of others. A third component would consist of interactive networks containing three or more individuals signalling to one another as well as eavesdropping on nearby interaction.

It's likely that the birds are gathering for a combination of reasons. They may be looking for food to break their nightly fast – although it's generally easier to see food sources when the light is a little brighter and the insects have come out in force. It may be a time for demonstrating lust and energy as the males start to engage in reproductive activities. It may also be a waking-up ritual – an injection of energy for the forthcoming day, like your morning coffee. It's especially fascinating that the singing begins minutes before any light actually appears. Nature awakes each day with its own mysterious alarm clock, and it awakes in chorus.

Just as dawn chorus singing is practised by many species of birds, vocal chatter is not uncommon in mammalian species. Numerous social groups of monkeys, such as baboons living in groups of 80 or more individuals, also indulge in daily breakfast chatter. Cows cluster together and moo together in the early hours of the morning. Roosters certainly let it be known that the sun has come up.

And there's a dusk chatter and vocalisation by nocturnal species such as bats, owls, frogs and toads. Since their daily cycles are shifted by roughly twelve hours, their dusk chorus may be functionally equivalent to the dawn chorus of other species. Bats engage in a lot of chatter just before flying out to forage for insects at dusk. The sounds and the echoes from the walls can be heard loudly and clearly if one stands in the entrance to a cave in the evening. This is despite the fact that much of the acoustic energy in the ultrasonic frequencies is beyond our hearing range. And of course frogs and toads living in the tropics or a subtropical climate are the greatest contributors to the dusk chorus, the males aggregating to advertise for females. Research has shown that fine-scale patterns of signal timing have a large influence on female choice. Females have a preference for leading, but not necessarily overlapping, auditory signals.

In some species, an entire group of birds may join in a chorus song to defend the territory. Groups of Australian magpies produce a remarkable chorus, varying from a quiet warbling to a loud caroling. White-crested laughing thrushes also sing in a group chorus. Each individual has its own phrase to contribute to the song: the result is like one bird singing.

★

The rhythms of nature explode in dance and song, creating further emotion and effect. When in harmony, all goes well. Discord feels wrong and incomplete. Dissonance disturbs. In the very deepest way, all of life responds to the rhythms around – by growing or dying.

11

Flirting, Courting and Coupling

The kiss originated when the first male reptile licked the first female reptile, implying in a subtle, complimentary way that she was as succulent as the small reptile he had for dinner the night before.

F. Scott Fitzgerald (1896–1940)

How do potential animal couples find and engage with each other? How are choices made? Besides marvelling at some of nature's ingenious and creative approaches to the mating game, it can be argued that *Homo sapiens* have a lot to learn. The inherent survival-of-the-species mandate creates limitless possibilities – with successes, failures and changes as the environment comes up with new requirements. For example, the male Anna's hummingbird has developed a way of emitting the same mating call both vocally and with his tail feathers. And for some animals procreating isn't a party, it's very hard work. The male Australian southern dibbler, a marsupial mouse, gets so exhausted during the few weeks of the mating season that he just lies down and dies. The male dibblers don't eat or sleep. They spend all of their time fighting with other males or mating with hard-won females. Each copulating session can last up to three hours, and afterwards the couple may remain together another nine to ten hours. Such stress soon collapses the male's immune system and within a few days he can develop bleeding ulcers and kidney failure; he is also vulnerable to a variety of infections and parasitic attacks. Meanwhile the female raises their offspring and probably lives for another couple of years to find (and exhaust) new mates and raise more litters. It's thought that the short life of the male aids the general survival of the species in a stark, inhospitable environment with little food.

The spring air is usually teeming with sensory cues and vibrations. The peeper frogs from Chapter 6 thaw out and hop around with gusto. It's the time for flirting and coupling. We define flirting as sending out signals of attraction or availability and evaluating the responses. Humans may flirt to spice up a conversation, make the day less boring, feel good about ourselves, or just because there's a sexual feeling in the air that stirs us for the moment. Often, that's the end of it. But for other animals – especially non-primates – flirting usually isn't an activity in and of itself. Fun as it may be – and we don't know for sure if it's fun for other animals or not – it's just the first step towards a serious mandate to keep the species going. That doesn't mean there aren't creative approaches. In fact, it could be argued that because non-human animals seem to be more attuned than humans to visual signals and information such as pheromones and electrical vibrations, flirting is even more creative and the next step of a relationship is more successful. Brief as it may be. And fatal for some.

Invertebrates and fish will also go to extraordinary lengths and behaviour in order to conquer the constraints of difficult surroundings. When asked about the sexual behaviour of fish, a noted marine biologist said: 'They do everything and sometimes at the same time.'

But it's the birds and their songs that stir many of us. Mating for birds is a never-ending party of song and dance and a flash of colour. And it may very well be that one of the best song-and-dance teams is composed of two Costa Rican long-tailed manakin males. University of Wyoming researchers headed by David McDonald filmed the phenomenal dancing duet of two male long-tailed manakins who cooperate as mentor and student to impress eligible females. The females will mate only with the alpha male – and although the females don't know the particular male, they have mapped the sites where the best dancing duets take place. After playing second fiddle to the alpha male, the beta male steps up to the alpha position eventually when his mentor 'retires'. While five years is the average waiting time for the beta male, it may take up to twenty years – but that's the deal with such a valued apprenticeship. The male long-tailed manakins all know each other, and through their social interaction they decide the mentorship duos and in which places to perfect their spectacular dancing displays before the waiting females and potential mates for the alpha males.

While most birds don't sing and dance at the same time, nearly all birds use sounds to call to their mates, their chicks or other birds. A lot of bird energy is devoted to singing; and while human enjoyment of singing birds may be readily apparent, it's probable that the birds also enjoy their own singing and that of others – the main purpose of birdsong being to mate and to protect territories and young ones. Although it's usually the male that sings to attract mates or defend territories, in some species, such as the alpine accentor, the female joins in as well. As Sir David Attenborough states in his documentary series on the lives of birds, territorial songs carry over long distances and convey detailed information about the location and the identity of the singer. Short intervals of silence in the song enable the singer to listen for replies, and to determine the location and distance of any possible rival.

The female black-headed grosbeak (a finch of the western USA) seems to use song to give her man a scare. Grosbeak parents take turns incubating the eggs. But if the male is late in returning to relieve the female from her duty, she sings a complex song imitating a male grosbeak. This may be to trick her mate into getting home quickly by giving the impression that a rival male is in his territory. Ravens do something similar. Members of a pair learn each other's calls, and when one partner is away or out of sight, the other will often send out its signature call to relocate it and prompt the absent mate's return. In some bird species, the female may join the male in singing to defend their mutual territory. Both sexes will sing a duet in alternation. Duetting is particularly common in birds that maintain year-round territories.

African shrikes are famous for their musical, repetitive duets – sounding like a cross between a bell and a horn. Each of the pair develops a unique duet 'part' which it uses to keep track of the other – whether in dense vegetation, to maintain their territory, or to synchronise their breeding cycles. Two shrike mates can become so synchronised in their singing that they sound like one. As mentioned in the preceding chapter, magpies in the Australian bush can also perform a precisely coordinated duet. One bird sends out a loud metallic 'tee-hee', immediately followed by the other's 'pee-o-wee, pee-o-wit'. Masters of harmony, the pair together sings one individual song.

Birds' evolutionary paths diverged from ours at or even just before the time of the dinosaurs. But just as humans wear colourful clothes for flirting and courting, so do birds. The peacock is a prime example. Peacocks can offer a dazzling display of fluorescent colours accompanied by dancing to attract the female peahens and mate with them. In contrast, the much more demurely coloured house sparrow mates for life and seems to know much more about establishing and maintaining relationships than we do. There is evident emotion in the songs and interactions of birds. They really do take mating seriously.

Before focusing further on sexual interactions between males and females of a variety of other species, we'll briefly look at other reproductive strategies such as forms of asexuality, reproducing without a partner – including parthenogenesis (virgin birth) and hermaphroditism (where an animal has both male and female sex organs). We'll also see that asexuality in animals can occur in combination with sexuality. Reproduction can get very complicated, and is influenced by environmental conditions.

Virgin birth

Parthenogenesis occurs when a female's eggs develop into babies without being fertilised. This can result in a form of cloning where the offspring is virtually genetically identical to the mother – or at other times slightly different from the mother and from other offspring. In the latter case, the meiosis (cell division) occurs in such a way as to mix up the genetic material ever so slightly. Many insects and other invertebrates such as water fleas come from unfertilised eggs. Some aphids do, too.

In optimal conditions, cloning can allow a species to create a large number of offspring in a short period of time. But many asexual animals can switch over to sexual reproduction during harsh or unpredictable times. The tiny aphid can reproduce asexually or sexually depending on the time of the year. During the spring when conditions are good, the female, through parthenogenesis, can give birth by unfertilised eggs to many other females that are born pregnant, and the aphid line multiplies quickly and prolifically. However, when conditions are not so good and host plants die, some females can turn into males that mate with females; they produce and fertilise eggs that are able to survive the winter cold or other crises such as a scarcity of food.

Other animals have been found to reproduce by parthenogenesis when they can't reproduce in their usual sexual manner. This has been proved over the years in some bony fish, amphibians, reptiles and birds. A virgin birth recently happened in Britain's Chester Zoo. A female komodo dragon that had no males around produced offspring through unfertilised eggs. This was the first time that parthenogenesis was documented in this Indonesian species. Obviously, this dragon was tired of waiting to be 'set up' with a proper mate.

Also surprising to scientists have been the recently documented virgin births in captivity of several types of sharks – the hammerhead, the bonnethead (also known as the shovelhead) and the blacktip. Scientists aren't sure whether such incidents occurred because the females had no males to accompany them in the different zoos and aquariums where they are housed, or if this in fact happens more often in nature than previously assumed. Shark scientist Dr Demian Chapman of the Institute for Ocean Conservation at Stony Brook University thinks that this very well might be 'something female sharks of many species can do on occasion'.

Hermaphrodites and sequential hermaphrodites

Hermaphrodites have both male and female sex organs. This gives them options that other animals don't have. In the absence of other members of the species, they can mate with themselves. If other members are around, it's not so difficult to find a mate. The following examples of hermaphrodites go about reproduction in different ways.

Giant clams that are found in the Indian and the southwest Pacific oceans can't move, so they can't roam around to find mates. Problem solved: they're hermaphrodites and can fertilise themselves if necessary. In the middle of summer they release eggs and sperm into the water – hoping to blend with other giant clams' eggs and sperm. They try to avoid fertilising their own eggs by spewing out the sperm at a different time – usually before the eggs. Also, the clams are male during their immature years and can send out only sperm. In later years they also develop female organs and eggs. And these giant clams are thought to be able to live almost 200 years. As complex as their reproductive process sounds, it certainly seems to go on for many, many years.

Another hermaphrodite, the mangrove killfish, lives in South American and southern US coastal swamps that can either dry up or become so toxic that the fish has to find refuge in the mud or by flipping and jumping across land. Amazingly, its skin and gills change so the killfish can breathe air and survive out of the water for as long as ten weeks. But this fish often leads a very lonely existence. If it can't find a mate, it fertilises itself by producing both eggs and sperm that mix together in an organ called an ovotestis. The fertilised eggs are then placed out of the water and hook onto plants or are dispersed by rain, tides or wind.

But life is more social for sea snails and other types of snail. As hermaphrodites, every member of their species is a potential mate. Snails pair up and carry out lengthy courtships that end when each inserts its long penis into the other to deposit sperm. Some snails don't want paternity competition, so they bite the penis off their partner after mating is finished.

There are also sequential hermaphrodites that produce egg and sperm, but not at the same time. Some produce only sperm when they are younger and later only eggs, while others produce only eggs in their early years and change to only sperm production in later life. All of this variety seems to ensure continuation of the species in a multitude of conditions.

Let's now look at sexual interactions (those with separate males and females) and their results. Scientists believe that sexual reproduction has gone on for at least 380 million years – more than three million years longer than previously thought. After mating, some couples are together only briefly, while others mate for life. In the final part of this chapter, we'll show why some higher-level species have found social monogamy to be the best strategy for survival.

The superorganism

Twentieth-century science focused mainly on reductive research, breaking a system (or a human body) into many parts and processes. In this century we have elevated our focus to patterns and systems, with a recognition that taking into account the context of any physical part is essential in understanding what's really occurring.

Since social insects, such as bees and ants, form and function within exceedingly complicated colonies that have ensured their survival for millions of years, they set the stage for conceptualising sexual and all other behaviour within a larger framework – their unity and complex biological organisation tell us much about our own species and ecology. How does mating feature in these and all complex systems? Those who have researched human sexuality agree that when we are struck with feelings of love and attraction, we rarely have a whit of concern about the social and ecological influences and consequences. Understanding of the broader context isn't a part of our thinking at such a time of passion and biochemically-induced psychosis. However, understanding the patterns and complex systems involved in perpetuating and maintaining individual species, as well as life itself, now has scientists looking at social behaviour at many levels.

Queens of the leafcutter ant and of the Florida harvester ant mate many times. The workers of both species are not only offspring of their mother but also of multiple males. As a result, the ants have many genetic predispositions and are able to respond to specific environmental requirements by expressing (activating from dormancy) the relevant genes. The same has been shown in the domestic honey bee. The results of multiple parental combinations mean that the colony can react more rapidly and flexibly, for example to the need for a customised labour system that carries out work for a specific time and place. Flexibility of the community and its offspring, as well as pragmatic and timely responses to environmental changes, is a plus for any species.

The quantity of offspring has long been considered important for survival of a species. As scientific knowledge increases, however, more emphasis is placed on the importance of quality – the physical, psychological and social strength of an individual and species. The health and vigour of an individual and population are even more essential than numbers, especially as we go up the scale of individual cognitive and anatomical complexity to humans.

Couples: finding, choosing, engaging

Finding a mate in a close-knit social compound is not as difficult as advertising for one at a distance. Usually it's the male that tries to contact and impress eligible females, through bright colours – like the male purple emperor butterfly whose bright purple coloration is visible to the female only from certain angles and in certain light conditions – and other impressive physical attributes (such as the acrobatic great crested newt that performs handstands, and the red-headed rock agama, a lizard that does push-ups), provocative sounds, and seductive movements. Of course, irresistible pheromones are always a part of the mate-catching formula, and the female plays an equal role in this.

This appears to be true for humans as well. Although we aren't as consciously aware of pheromones as other animals may be, studies at Stanford University showed that the smell of a person ranked at the top of the list of attractive attributes. Pheromones are also important in sustaining a relationship, as evidenced in Napoleon Bonaparte's famous message to his wife Josephine: 'I'm coming home in three days. Don't bathe.'

Emitting sounds to attract a mate crosses almost all species lines, from the male mole crickets that rub their wings together in a burrow that amplifies the 'song', to human crooners and rock stars. Cicadas play their tymbals, a pair of organs on their abdomens that, like flat metal cymbals, can be banged against each other to create one of the loudest noises in the natural world. Male frogs that croak their interest to available females are known to purposefully change back and forth from higher to lower frequencies to seem larger or more interesting, or to trick rivals or predators.

Brown University researchers recently described a pattern of stuttering communication techniques in male bullfrogs that had never been reported in the scientific literature. The scientists recorded 2,536 calls from 32 male bullfrogs in natural chorus and analysed the croaks in each call as well as the number of stutters in each croak. They found that the stutters were not a result of fatigue or because a frog was running out of breath, nor were they a part of aggressive or territorial communication. The researchers concluded that the stutters seemed to have the purpose of extending the length of an individual call, and since they are produced more frequently during certain times in the breeding season, they may very well be a part of attracting a mate.

Bullfrog males change the pitch and amplitude level of their croaks, depending on whether they want to challenge rival suitors or seduce a willing female frog. Male moose moaning calls can be heard by females as far as 2 miles (3.2 km) away. And the roar of the male elephant seal announcing his fight for a mate travels an even greater distance.

The plainfin midshipman fish use two strategies to get females to their nests in the intertidal area of the North American Pacific coast from Mexico to Alaska. First the males call out. And the midshipman fish really can call – loudly and intensely. They send out a reverberating humming sound interspersed with grunting, whistling and growling. It's hard for anyone in the vicinity to miss these night-time mating invitations. Next the males 'shine their lights'. When a female comes close, the male will arch his head back to boldly expose the luminescent spots on the underside of his chin, and she's hooked.

'Light shows' of various types can be part of advertising as well as creating an engaging ambience. The beautiful male ornate jumping spiders have light-reflecting 'fluorescent' scales on different parts of their bodies, especially their forelegs. These brilliant scales shine in the sunlight as the male jumping spider performs his intricate courtship dance.

Summer night fireflies broadcast their spectacular light shows filled with drama and intrigue. Ultimately, the female chooses from hundreds of sparkling suitors – settling on the love light of one male. The light is produced by an enzyme in the firefly's tail that creates a chemical reaction. Of the more than 2,000 firefly species throughout the world, some make flashes as adults, others only glow, and still others don't give off any light at all. Tufts University ecologist Sara Lewis has been studying these beautiful light displays and even participating in them with her specially designed penlight. When she clicks it twice in perfect *Photinus greeni* – double pulses, then three seconds of darkness, then two pulses – a female *Photinus greeni* flashes back from the meadow grasses of the eastern Massachusetts farm field. In this one small area Lewis has found six firefly species, each with its own pattern of lights. The *Photinus ignitus* male gives off single pulses separated by five-second delays. The mating signals are intricate and precise, with thousands of males light-signalling in the air and the few females responding from the grasses below. Each female looks for the light pattern from a male of her species – then flashes a quick signal of her own. A firefly female may zone in on ten or more males in an evening – carrying on several light conversations at a time. Finally she decides on the Mr Right that she discerns has the best nuptial gift, as scientists call it – packages of proteins to be injected with sperm. Perhaps one of the most beautiful light shows is that performed by the *Pyractomena angulata.* According to Lewis, their flash pattern is 'like a flickering orange rain'. So while we humans have our fireworks and other light displays for special occasions, nature celebrates in even finer grandeur every summer night.

Other suitors, including humans, are thought to have especially active cryptochromes during a full moon, and that can lead to more romance. Many animals, from coral to humans, possess these light-sensitive genes that operate as part of our circadian system and adjust our cycles and tune us into rhythms all around us. In coral, these genes set off activity in ancient proto-eyes. Other animals have light-sensitivity in their retina and other parts of their eyes and bodies. Australian and Israeli researchers found that cryptochromes are responsible for triggering the annual mass spawning of Great Barrier Reef corals that follows a full moon.

In most animal groups the female often chooses the mating situation. And researchers have found that what a female bird may want in a mate can change from year to year. Alexis Chaine of France's Laboratory of Evolution and Biological Diversity and Bruce Lyon of the University of California studied lark buntings for five years. They found that each year different traits were associated with a 'fit' male. For example, having a large body and a lot of great rump feathers were signs of reproductive sex one year, then considered marks of reproductive failure the next year. Much of this seemed to be based on the changes in predators, climate and food. It seemed that each year female buntings were able to intuit what traits would be best for having the healthiest babies based on environmental changes.

In some animal groups, rough males may try to dominate the female's choice or aggressively change her mind. Along with good looks, wit, charm and intelligence can be trump cards in primate circles. But the female must be physically ready. While it's thought that females of all species have an awareness of this readiness to mate, a number of male animals check it out for themselves.

Snakes, giraffes, elephants and other animals have a way to test if a female is fertile. It's called Jacobson's organ. Jacobson's organ is a patch of specialised skin at the roof of the mouth or base of the nasal cavity of many vertebrates. This vomeronasal organ is very sensitive to airborne molecules of scents. Snakes and lizards use their organ to sense predators as they flick their tongues back and forth. Mammals use this organ mostly to sense pheromones. Such animals as lions, tigers, zebra and giraffes taste the air for pheromones by drawing their upper lip back in what's called a flehmen response and allowing air to flow over the Jacobson's organ. Sometimes males can pick up the airborne molecules from whiffs of pheromones emanating from the female to see if she's ovulating. Often they check her out even more closely. In the case of giraffes, the male will follow a likely female and nuzzle her genitals, which causes her to urinate. The male then bends down to catch some of the urine on his muzzle. Then he moves his head back up and twists his long tongue around, moving his mouth and lips in a grimace, Then he bares his teeth. The female stops to let him check her, then walks away. If she's fertile, he follows her and attempts to mount her. If she's not fertile, he leaves and goes on to search for a female who is.

For humans, not only is a person's scent detected through the air, but kissing probably gives even more biochemical data, as it occurs in the mouth and nasal area. Other primates kiss as well. In fact, bonobos, related to chimpanzees and native to the Congo, have been found to engage in French kissing.

Other mammals have also demonstrated kissing behaviour. Moustached bats frequently emit both affiliative whistling and harsher, aggressive sounds. Close observations have shown that friendly bat whistles are frequently accompanied by a lot of touching and 'kissing' and sometimes mouth-to-genital interactions between a male and a female. Bats also hug each other with their folded wings. Kissing usually consists of a brief contact between the mouths of two bats and may be accompanied by a quick lick of the significant other's lips as well. In contrast, the harsh broadband sounds are associated with boxing, nipping and aggressive biting. Fighting ensues in different situations. Sometimes, one bat may even take a big swing and whack another bat, knocking it off its hanging position. In one scenario, a male bat may engage in promiscuous hugging of females 'owned' by a dominant male. The dominant male immediately lets out a harsh call that is usually sufficient to turn the sneaky male away. The dominant male then produces whistling sounds and makes genital-to-oral contact with his females to reaffirm their bond.

Dogs also sometimes greet each other in what might be considered a 'kiss'. Once again, coming close to the breath and nasal and oral smells of each other can give a great deal of data – even to humans. And whether or not it can be considered actually 'kissing', many other animals lick or nuzzle each other in ways that pick up the sexual receptivity of the other.

The similarities between animals and humans even extend to specific mating practices. For a long time it was thought that only humans mated facing each other, in what is commonly known as 'the missionary position'. However, some apes and chimpanzees have recently been observed mating in this position as well.

In our study of other animals we are also able to learn more about ourselves. Nowhere is that more compelling in contemporary science than in the biochemicals of bonding, of enduring relationships and of love. After we get into a relationship, particular experiences, biochemicals and genetic interactions influence our ability to stay together. By studying those animals that mate for life, for example, we have found certain biochemicals in their systems that are not as abundant in animals that do not create such long-term bonds. And in this century, scientists have realised that it's not just a matter of somehow being born with the 'right genes'.

Scientists researching human mother-and-baby bonding discovered significant amounts of oxytocin in each. Oxytocin is a pleasure-inducing neurotransmitter that generates a sense of satisfying calm and closeness. In a strong interactive cycle, mother and baby respond to feelings of love by producing oxytocin, which in turn makes them feel even closer and so creates a type of positive feedback loop. Ever-increasing levels of oxytocin change not only which genes are expressed, but potentially also the actual genes themselves. In recent studies on rats, oxytocin has been found to alter the DNA itself. Once thought to be an unchangeable map of life that we are each born with, scientists now know that our DNA does indeed alter with our experiences and the consequent electrical and chemical changes that our experiences induce. For example, according to McGill University's Michael Meaney, these early environmental experiences – like a mother's care or lack of it – can induce chemical changes on the DNA strand.

We aren't limited by, nor have to be hostages to, our DNA. Every time we have a feeling of anger, we produce billions of molecules in our system. A feeling of love does the same. As Stanford University's David Spiegel likes to say: 'We all have our own little drug factories in our heads.' Repetition of certain emotions, actions and events creates cascades of chemical and electrical changes that RNA, our messenger molecules, take back and forth from the genes and cells, and which eventually affect our genetic map – our DNA.* Therefore, we not only change, we pass on that

* It has been known since 1961 that when a neuron is stimulated, it produces and spews out large amounts of RNA (ribonucleic acid) – the genetic messenger, modulating and activating molecule. Over the last decade, scientists have discovered increasingly different types of RNA, including those that interfere with commands given by genes, those that prompt expressions (actions) from certain genes depending on requirements coming from the environment or internally within a person's body, and those that configure different proteins and processes that are thought to actually change an individual's data.

change to our offspring. We have known since Darwin's studies of the genetic changes in the bills of finches that new situations and requirements change our genetic patterns. We have known of mutations – changes in genetic material caused by environmental or other pressures. But we didn't realise until recent epigenetic research (this is the study of the influence of the environment on genes and DNA) how very malleable our genetic proteins and molecules really are.

Coupling for life – or at least until the children leave the nest

Oh, the joys of being monogamous. Not having to keep fighting off rivals. Not having to go through tiring and often frustrating rituals to get a mate. Not getting sent off before the children are born. Most birds stay together for the year of producing and tending to babies. Numerous breeds of parrot and pigeon couples are known to stay together for many years. The royal albatross, thought to live for up to 80 years, mates for life. Ring-necked doves also mate for life, as do some species of penguins.

But it's the prairie vole of southern Illinois that has become most famous in scientific circles. Even compared to other neighbouring voles, the prairie voles (*Microtus ochrogaster*) are prototypes of monogamy. The male vole stays the rest of his life with the female vole with whom he loses his virginity. There's a practicality to staying with one mate. In a hostile environment with many predators, there aren't that many opportunities to find mates. And fighting with rivals or having sex with different mates would diminish the energy necessary for survival and raising a family. Then there's the biochemicals part of the story.

Thomas Insel's pioneering research found that in contrast to other voles, prairie vole males have higher levels of oxytocin – which, as we saw earlier, is known for its bonding effect and 'love feeling'. They also have heightened levels of vasopressin – associated with good feelings and memory – and dopamine, associated with pleasurable and rewarding feelings. Researchers are also studying the prairie voles to see what effect their social monogamy has on their hearts, the belief being that it's associated with healthier hearts. Whether these biochemicals are the source or product of monogamy isn't entirely certain, although it's likely that, as in the strengthening bond between human mothers and their infants, they are in fact both.

There is a reason to look more closely at the male prairie vole's level of another biochemical – the neurotransmitter serotonin. People with depression and anxiety have often been found to have low serotonin levels. And a recent study of swarming in migratory animals concludes that animals with low serotonin levels interfere with natural swarming behaviour. Most of all, scientists have shown how powerful these biochemicals are – how they facilitate our experiences and how we create more of them as a result of our experiences. Humans also benefit and couple as a result of this bonding and loving cycle, as Daniel Goleman writes in his book *Social Intelligence.* In the prelude to love-making, oxytocin levels soar in a man's brain, along with vasopressin – a hormone closely related to oxytocin that can also play a strong role in bonding. In both men and women, oxytocin generates loving and pleasurable feelings, and is released in large doses during orgasm. Feelings of love and a warm afterglow ensue. Oxytocin secretions continue to be strong, especially during 'afterplay' – the time of cuddling and continued closeness. According to Goleman, 'the oxytocin produced by orgasm also boosts memory, again imprinting in the mind's eye a lover's fond figure'.

And so the cycle continues. A good experience creates more 'loving' and 'pleasurable' biochemicals as well as good memories of the person with whom we shared such an experience.

Epilogue

'Human Nature' Reconsidered

What a piece of work is a man! How noble in reason! … in apprehension, how like a god!

William Shakespeare, *Hamlet* (1601)

'Are human beings in some way different, unique in creation in the eyes of God? Are we … special,' asked James Trefil in his 1997 book *Are We Unique?*. What is most compelling is not Trefil's question, which has been asked throughout history, but the scientific data he had to draw upon in his argument, and the difference in the vast research available to scientists little more than a decade later. Most of the scientific evidence in this book wasn't available to Trefil. So what we can consider to be specifically 'human nature' is an ongoing and ever-changing question, not only because there's still so much we don't know about other animals, but because it has been shown that all forms of life are changing genetically much more rapidly than we could have imagined even a year ago!

Yet, Trefil's question is valid, because an enormous portion of our behaviour and decision-making is based on the assumptions we have about 'human nature'. It has long been thought that humans are the only animals who possess a soul, who have a sense of consciousness, spirit and self-reflection. In the 20th century, scientists – especially psychologists – surmised that non-humans including 'lower' primates lacked the capacity to problem-solve, to plan for and predict future events, and to cooperate with others to accomplish a task. Other attributes we considered uniquely human until recently were: the ability to abstract, to create from such abstractions, and to accomplish certain memory tasks including the ability to hold in memory and to retrieve certain data in order to build and be creative. (All that we saw animals do that could be attributed to such skills was written off as 'instinct' – not thinking.)

We also suspected that it was distinctly 'human nature' to show compassion, deep love, and altruism. We weren't even sure if other animals were capable of some of our baser qualities, such as jealousy or revenge – or making weapons, or premeditated murder. And some scientists still wonder if 'instinct' trumps 'true learning' abilities and even the urge to teach.

Recent scientific data have upended many of our assumptions. Other primates, fish, birds, dogs, cats, elephants, hyenas, bats, mice and other species have many ways of living their lives that do indeed make us reconsider 'human nature'. Scientists have not yet gathered conclusive evidence as to whether non-human animals also have the need to feel superior to others – although it appears that other mammals indeed share this competitive attribute, as well as jealousy and even revenge, just as they share our finer qualities of caring and empathy. That other animals feel and express emotions seems a logical assumption. How and why can differ. None of us, because of our individual set of experiences which then influence our memories and future perceptions, can deeply know what another human experiences, nor can we be certain of other animals. At the same time, recent studies have confirmed a number of cases of altruistic behaviour in birds and mammals; and we can be open to appreciating that elephants may very well have a sense of self, as evidenced in psychological experiments. But when we hear dogs grieve for the animals and people they love with a low and mournful moan, and see that even more than humans, they have the capacity to know when a human has passed from this life, or when we view a deeply touching video showing distraught elephants coping with their own losses, we know first-hand the profound scope of nature.

The findings reported in this book rest on the hard work of researchers and scientists, many of whom have dedicated their careers to probing the secret lives of animals and the processes inside their brains, from the level of quantum events to molecules, to cells, to actual behaviour. These findings show that our relationship with animals is more of a lateral than a hierarchical one. Our abilities to perceive, think, feel, sing, dance, giggle and problem-solve emerge from and are shared by many other creatures.

We've also shown in this book certain animal features that humans either no longer possess or don't allow into our conscious recognition, arguably to our detriment. Most of all, ongoing scientific discoveries continue to challenge us to reconsider what we may take for granted as 'human nature'.

In his recent book, *The Wachula Woods Accord*, Charles Siebert writes: 'The degree to which we humans will finally stop abusing other creatures, and for that matter, one another, will ultimately be measured by the degree to which we come to understand how integral a part of us all other creatures actually are.' Our goal in this book has been to improve that understanding in those who yearn for it, as well as to stimulate others who have not yet begun to view the animals' inner worlds as a continuum of our inner worlds to begin to do so and to enjoy the animal world in this new light. We sincerely hope that this new understanding and perspective will improve the quality of life of all creatures and especially our own.

Notes, References, and Further Reading

Introduction

Kanwal, J.S. (2009), 'Audiovocal Communication in Bats', in Squire, L.R. (ed.), *Encyclopedia of Neuroscience*, Vol. 1, pp. 681–90.

Balcombe, J.P. (1990), 'Vocal recognition of pups by mother Mexican free-tailed bats: do pups recognize their mothers?', *Animal Behaviour*, 39, 980–986.

Knörnschild, M., Behr, O., and von Helversen, O. (2006), 'Babbling behavior in the sac-winged bat (*Saccopteryx bilineata*)', *Naturwissenschaften*, DOI 10.1007/s00114-006-0127-9, URL http://dx.doi.org/10.1007/s00114-006-0127-9, Oxford: Academic Press, pp. 1–4.

Chapter 1: A Supercharged World

Some of the latest research on the human 'electric brain' and articles by Karl Pribram:

Pribram, K.H. (2004), 'Brain and Mathematics', Chapter 12 in Globus, G., Pribram, K., Vitiello, G. (eds), *Brain and Being: At the Boundary between Science, Philosophy, Language and the Arts.*

Pribram, K.H. (2004), 'Consciousness Reassessed', *Mind and Matter*, Vol. 2, No. 1, pp. 7–34.

For a magical video demonstration of the magnetic waves that surround us, see: http://www.semiconductorfilms.com/root/Magnetic_Movie/Magnetic.htm

Throughout this book – especially in Chapters 1, 2 and 11 – we have relied heavily on a recent extraordinarily beautiful compilation of photographs and animal data by the American Museum of Natural History, DK Publishing, 2008. This 500-page compendium of animal research, entitled *Animal Life*, could easily be recommended as a visual and companion reference to our work here. Educated at Bristol, the editor-in-chief, Charlotte Uhlenbroek, brought together an impressive group of contributors and consultants, many from the UK. We will begin each chapter by designating which pages of *Animal Life* we have used as resources, abbreviating each citation as AL, as below:

AL: electricity and magnetism – pp. 119, 130–1, 218, 234, 276, 443, 462, 464; short-beaked echidna – pp. 70, 130, 394; duck-billed platypus – pp. 70, 130, 234.

See also for electro-sensitivity in the bill of the duck-billed platypus: Manger, P. and Pettigrew, J., 'Electroreception and the feeding behaviour of platypus', *Philosophical Transactions: Biological Sciences*, 347 (1322), 359–381.

In the very first step of sensation, all physical signals are transduced or converted into minuscule voltage changes in the nervous system. It's as if energy in the outside world, be it in the form of electromagnetic fields, light, sounds or smells, is giving us tiny shocks all the time. Yet it's not the little shocks we sense but the real objects and the meaning they convey. All animals use this information and adapt their behaviour so that they can navigate, survive and thrive in their individual environments.

A biological 'system' is composed of receptors that detect the molecules and nerve cells that carry the information to the brain. In the brain, there are well-defined roadways that transmit information to specific target areas, where the information can be physically processed, just like a raw product may be processed over an assembly line to create a marketable product in the manufacturing industry. The processed information is like food for brain cells, which manage new incoming information and trigger actions to guide an animal, e.g. towards its food source.

Electroreception, sometimes called electroception, is the biological ability to perceive electrical impulses. For further details, see http://en.wikipedia.org/wiki/Electroreception. Carl D. Hopkins is actively engaged in the study of electroception and electrocommunication. Visit his website at http://www.nbb.cornell.edu/neurobio/hopkins/research.htm to find out more about his research.

Menon, C., *Science Direct – Acta Astronautica*, Vol. 64, Issue 4, February 2009, pp. 395–495.

Weiss, R., 'Mind over Matter: Brain Waves Guide a Cursor's Path', *Washington Post*, 13 December 2004.

BBC News/Health, 'Brain Chip Reads Man's Thoughts', 31 March 2005.

To discharge the electrocytes at the correct time, the electric eel uses its pacemaker nucleus, a nucleus of pacemaker neurons. When an electric eel spots its prey, the pacemaker neurons fire and acetylcholine, a neurotransmitter, is subsequently released from electromotor neurons to the electrocytes, resulting in an electric organ discharge (EOD).

Kalmijn, A.J. (1982), *Science*, 218, 916–918.

Bell, C.C. and Grant, K. (1989), *The Journal of Neuroscience*, 9, 1029–1044.

Zakon, H.H. and Dunlap, K.D. (1999), *Brain, Behavior and Evolution*, 54, 61–69.

Bennett, M.V.L. (1971), 'Electric Organs', in Hoar, W.S. and Randall, D.J. (eds), *Fish Physiology*, Vol. 5, Chapter 10.

Belbenoit, P., Moller, P., Serrier, J. and Push, S. (1979), 'Ethological observations on the electric organ discharge behaviour of the electric catfish, *Malapterurus electricus* (Pisces)', *Behavioral Ecology and Sociobiology*, 4, 321–330.

Carr, C.E., Maler, L. and Taylor, B. (1986), 'A time-comparison circuit in the electric fish midbrain. II. Functional morphology', *The Journal of Neuroscience*, 6, 1372–1383.

Heiligenberg, W., Baker, C. and Matsubara, J. (1978), 'The jamming avoidance response in Eigenmannia revisited: the structure of a neuronal democracy', *The Journal of Comparative Physiology [A]*, 127, 267–286.

Heiligenberg, W., Finger, T., Matsubara, J. and Carr, C. (1981), 'Input to the medullary pacemaker nucleus in the weakly electric fish, eigenmannia (*Sternopygidae, gymnotiformes*)', *Brain Research*, 211, 418–423.

Kawasaki, M. and Heiligenberg, W. (1989), 'Distinct mechanisms of modulation in a neuronal oscillator generate different social signals in the electric fish hypopomus', *The Journal of Comparative Physiology [A]*, 165, 731–741.

Wu, C.H. (1984), 'Electric fish and the discovery of animal electricity', *American Scientist*, 72 (Nov–Dec), 598–606.

The coordinates of the Bermuda triangle vary with the author, but this trapezoidal area includes the Atlantic coast of Florida; San Juan, Puerto Rico; and the mid-Atlantic island of Bermuda.

Lohmann, K.J. and Lohmann, C.M.F. (1996), *Nature*, 380, 59–61.

Lohmann, K.J. and Johnsen, S. (2000), *Trends in Neurosciences*, 23, 153–159.

Footnote on p. 14: Philine Feulner in a *New Scientist* article, 26 November 2008, 'Electric fish prefers sexual charge of its own species', *Biology Letters*, DOI: 10.1098/rsbl.2008.0566.

Diego-Rasillo, F.J. and Phillips, J.B. (2007), 'Magnetic compass orientation in larval Iberian green frogs, *Pelophylaxperezi*', *Ethology*, 113, 1–6 (pdf).

Muheim, R., Phillips, J.B. and Akesson, S. (2006), 'Polarized light cues underlie compass calibration in migratory songbird', *Science*, 313, 837–839 (pdf).

Freake, M.J. and Phillips, J.B. (2005), 'Light-dependent shift in bullfrog tadpole magnetic compass orientation: evidence for a common magnetoreception mechanism in anuran and urodele amphibians', *Ethology*, 111, 241–254 (pdf).

Diego-Rasillo, F. J., Luengo, R.M. and Phillips, J.B. (2005), 'Magnetic compass mediates nocturnal homing by the alpine newt, *Triturus alpestris*', *Behavioral Ecology and Sociobiology*, 58, 361–365 (pdf).

Wente, W.H. and Phillips, J.B. (2005), 'Microhabitat selection by the Pacific treefrog, *Hyla regilla*', *Animal Behaviour*, 70, 279–287 (pdf).

Johnsen, S. and Lohmann, K.J. (2008), 'Magnetoreception in animals', *Physics Today*, 61(3), 29–35; download pdf: http://www.biology.duke.edu/johnsenlab/pdfs/pubs/physics%20today.pdf

Ritz, T., Adem, S. and Schulten, K. (2000), 'A model for photoreceptor-based magnetoreception in birds', *Biophysical Journal*, 78, 707–718.

Canfield, J.M., Belford, R.L., Debrunner, P.G. and Schulten, K. (1995), 'A perturbation treatment of oscillating magnetic fields in the radical pair mechanism using the Liouville equation', *Chemical Physics*, 195, 59–69.

Schulten, K. (1986), 'Magnetic field effects on radical pair processes in chemistry and biology', in Bernhard, J.H. (ed.), *Biological Effects of Static and Extremely Low Frequency Magnetic Fields*, MMV Medizin Verlag, Munich, pp. 133–40.

Birds and magnetic fields: 'Homing pigeons use earth's magnetic field', an article in the *Daily Telegraph*, 24 June 2009, relates the latest research from the University of Auckland, just published in the journal *Proceedings of the Royal Society B*, which offers further evidence that birds use the magnetic field to navigate and that magnetic particles in bird beaks act like compasses. Researchers believe such is not unique to the homing pigeons; it could be universal among all birds and present in other animals influenced by the earth's magnetic field.

Wiltschko, R. and Wiltschko, W. (1996), *Animal Behavior: Magnetic Orientation In Animals*, *Zoophysiology*, Vol. 33, Berlin and New York: Springer-Verlag.

Bats famously orient at night by echolocation, but this works over only a short range, and little is known about how they navigate over longer distances. It was recently shown by Holland and colleagues that the homing behaviour of *Eptesicus fuscus*, known as the big brown bat, can be altered by artificially shifting the earth's magnetic field, indicating that these bats rely on a magnetic compass to return to their home roost. This finding adds to the impressive array of sensory abilities possessed by this animal for navigation in the dark. For details, see: Holland, R.A., Thorup, K., Vonhof, M.J., Cochran, W.W. and Wikelski, M. (2006), 'Bat orientation using Earth's magnetic field', *Nature*, 444, 702, doi:10.1038/444702a.

Chapter 2: Good Vibrations

AL: vibrations – pp. 120–21.

Ovidu Lipan's quote is from his article in the 25 January 2008 edition of the journal *Science*.

The story of Tesla was taken from Cheney, M. (2001), *Tesla: Man out of Time*, Simon & Schuster (quotation from out-of-print Dell edition, 1981).

Theoretically, substrate-borne vibrations can be detected and processed by the somatosensory system, the auditory system, or both. The high resemblance in latency, shape and duration of the responses evoked by the vibratory stimuli and by high-intensity airborne clicks, the almost complete elimination of these responses by masking noise or by deafening the animal, all provide solid evidence for the primary role of the auditory system in the processing of the vibratory signals.

Wollberg, Z. (2006), in Kanwal, J.S. and Ehret, G. (eds), *Behavior and Neurodynamics for Auditory Communication*, Cambridge University Press, pp. 36–56.

Rado, R., Himelfarb, M., Arensburg, B., Terkel, J. and Wollberg, Z. (1989), *Hearing Research*, 41, 23–29.

Mole rat seismic signalling: the blind mole rat is from Wollberg, Z., Rado, R. and Sadka, R. (2006), 'The Blind Mole Rat: An Example of Seismic Communication via Acoustic Channels', in Kanwal, J.S. and Ehret, G. (eds), *Behavior and Neurodynamics for Auditory Communication*, pp. 36–56. The Cape mole rat: Narrins, P.M. et al. (1992), 'Seismic signal transmission between burrows of the Cape mole-rat', *Journal of Comparative Physiology*, 209, 4238.

A good resource is: Pikovsky, A., Rosenblum, M. and Kurths, J. (2003), *Synchronization*, Cambridge University Press.

Hill, Peggy (2008), *Vibrational Communication in Animals*, Harvard University Press.

Makris, N. et al. (2009), 'Critical Population Density Triggers Rapid Formation of Vast Oceanic Fish Shoals', *Science*, 323, 1734–1737.

Prior to Makris' research, a number of other studies have been done on swarming and other collective behaviour, including the following which also discusses the importance of density in transition of disorder to an orderly swarming: Buhl, J. and Sumpter, D. (2006), 'From Disorder to Order in Marching Locusts', *Science*, 312, 1402–1405.

In the case of locusts, Michael Anstey at the University of Oxford and Stephen Rogers at the University of Cambridge, and colleagues at the University of Sydney, showed that the neurotransmitter serotonin is necessary for bringing the locusts closer together to reach the threshold density level for swarming: Anstey, M. et al. (2009), 'Serotonin Mediates Behavioral Gregarization Underlying Swarm Formation in Desert Locusts', *Science*, 323, 627–629. Earlier in the same issue of *Science*, Stevenson, P.A., 'The Key to Pandora's box', 323, 594–595 writes about Anstey's study. The importance of serotonin in locust swarming and other behaviour is mentioned in Chapter 11.

Showing examples of self-organising: a good description of the study of green-headed ants by Audrey Dussutour and Stephen Simpson of the University of Sydney in Australia is the article, 'Ants Adjust Foraging so the Colony Eats Right', 'Science', *New York Times*, 21 April 2009; G.W. Baxter of Penn State, Erie and others are using mathematical models to describe the foraging of small black ants as well as the collective and swarming behaviour of many other species.

Very good examples of self-organisation are given throughout Holldobler, Bert and Wilson, E.O. (2009), *The Superorganism*, Norton – including the nest architecture quote found on page 473.

Creating entangled pairs of photons is difficult, but physicists from the UK and Japan have found a way to do it by passing ordinary photons through a novel optical filter. 'Photon Sieve Lights a Smooth Path to Entangled Quantum Weirdness' (2009), is a very readable article by Adrian Cho, *Science*, 323, 453; research: 483–485.

For J.G. Rarity: 'Ground to satellite secure exchange using quantum cryptography' (2002), *New Journal of Physics*, 4, 82, doi: 10.1088/1367-2630/4/1/382.

Rarity, J.G., Gorman, P. and Tapster, P.R., 'Free space quantum cryptography and satellite secure distribution', *Nanotechnology and Quantum Computing*, ref. no. 2000/140, IEE seminar on 02/2000.

For NIST: Jost, J.D., Home, J.P., Amini, J.M., Hanneke, D., Ozeri, R., Langer, C., Bollinger, J.J., Leibfried, D. and Wineland, D.J., (2009), 'Entangled mechanical oscillators', *Nature*, 459, 683–685, 4 June 2009, doi: 10.1038/nature08006, letter.

Chapter 3: Sounds for Tracking and Talking

The term echolocation was first coined by Donald Griffin in 1944 and refers to the use of sound reflections to localise objects and orient in the environment. For further details, see 'Echolocation' on p. 253 of *The MIT Encyclopedia of the Cognitive Sciences* (2001), by Robert Andrew Wilson and Frank C. Keil, MIT Press.

Sound intensity is measured in decibels of sound pressure level or dB SPL. The scale of sound pressure is based on human hearing sensitivity and ranges from 0 to 100 dB SPL. 0 dB SPL is the threshold level below which sounds become inaudible. Beyond 100 dB SPL, sounds start becoming painful. Between 120 to 140 dB SPL, sounds can cause permanent damage to a person's hearing ability and even result in a condition called tinnitus, where a person continuously hears an annoying (usually high-pitched) tone in the absence of any sound.

Clement, M.J., Dietz, N., Gupta, P., and Kanwal, J.S. (2006), in Kanwal, J.S. and Ehret, G. (eds), *Behavior and Neurodynamics for Auditory Communication*, Cambridge University Press, pp. 57–84.

Griffin, D.R. (1986), *Listening in the Dark*, Cornell University Press, pp. 96–119.

Kanwal, J.S., Matsumura, S., Ohlemiller, K. and Suga, N. (1994), 'Analysis of acoustic elements and syntax in communication sounds emitted by mustached bats', *Journal of the Acoustical Society of America*, 96, 1229–1254.

Kanwal, J.S. and Suga, N. (1995), 'Hemispheric asymmetry in the processing of calls in the auditory cortex of the mustached bat', *Journal of the Association for Research in Otolaryngology*, St Petersburg Beach, FL, p. 104.

Kanwal, J.S., Matsumura, S., Ohlemiller, K. and Suga, N. (1994), *Journal of the Acoustical Society of America*. 96, 1229–1254.

Leippert, D. (1994), *Ethology*, 98, 111–127.

Suga, N. (1990), 'Biosonar and neural computation in bats', *Scientific American*, 262, 60–68.

Suga, N. and O'Neill, W.E. (1979), 'Neural axis representing target range in the auditory cortex of the mustached bat', *Science*, 206, 351–353.

A dolphin has two dorsal bursa/phonic lip complexes, which can operate independently and simultaneously. The dolphin is able to generate clicks within its nasal sacs, situated behind the melon. The melon acts as a lens, which focuses the sound into a narrow beam that is projected in front of the animal. When the sound strikes an object, some of the energy of the soundwave is reflected back towards the dolphin.

Au, W.W.L. (1993), *The Sonar of Dolphins*, Springer.

http://www.cbmwc.org/education/echo.asp

May, J. (1990), *The Greenpeace Book of Dolphins*, Century Editions.

Payne, R.S. and McVay, S. (1971), 'Songs of humpback whales', *Science*, 173, 585–597.

Most recently, a third species of bats, Mexican free-tailed bats, have also been shown to sing (Bohn, K.M., Schmidt-French, B., Ma, S.T., Pollack, G.D. (2008), 'Syllable acoustics, temporal patterns and call composition vary with behavioral context in Mexican free-tailed bats', 561, *Journal of the Acoustical Society of America*, 124, 1838–1848, 562). Moreover, in a recent study,

Yovel et al. show that greater mouse-eared bats recognise the voice of conspecifics from multiple call characteristics in the echolocation pulse (Yovel, Y., Melcon, M.L., Franz, M.O., Denzinger, A., Schnitzler, H.-U. (2009), 'The Voice of Bats: How Greater Mouse-eared Bats Recognize Individuals Based on Their Echolocation Calls', PLoS *Computational Biology*, 5(6), e1000400. doi:10.1371/journal.pcbi.1000400).

Riskin, Daniel K., Bahlman, Joseph W., Hubel, Tatjana Y., Ratcliffe, John M., Kunz, Thomas H., Swartz, Sharon M. (2009), 'Bats go head-under-heels: the biomechanics of landing on a ceiling', *Journal of Experimental Biology*, 212, 945–953.

To learn more about bats and other organisms, go to: http://www.nbii.gov/portal/community/Communities/Plants,_Animals_&_Other_Organisms/

Chapter 4: Tasting and Touching

Sensitivity to touch is greatest around the lips, nipples, eyes and fingertips. Body hairs are also good sense organs. Keratin, found on the surface of the skin, has a piezoelectric effect: that is, whenever the epidermis is bent or folded, detectable electric discharges are produced which are picked up by the free nerve endings and translated in the brain.

Most of the smells that we can detect and describe (e.g. the smell of lilacs) are composed of a specific combination of dozens of different odour molecules. It is the function of the olfactory system to detect these various individual odour molecules and then to integrate these individual signals into a composite scent. Animals with sensitive olfactory systems can detect thousands of odorants and have over 1,000 genes encoding for olfactory receptor proteins. Humans have a poor sense of smell and have possibly only 500 genes for olfactory receptors. The olfactory system is a very primitive system that helps track prey and remember noxious stimuli. The receptor cells for olfaction are located in the olfactory mucosa in your nose, which has an area of about 5–10 cm^2 in humans and is much larger in some animals with an acute sense of smell. The olfactory receptor neurons send their axons through the cribriform plate of the ethmoid bone to terminate in the olfactory bulb.

Herrick, C.J. (1904), 'The organ and sense of taste in fishes', *Bulletin of the US Fish Commission*, 12, 237–272.

Herrick, C.J. (1919), 'A study of the vagal lobes and funicular nuclei of the brain of the codfish', *Journal of Comparative Neurology & Psychology*, 67–97.

Herrick, C.J. (1919), 'The tactile centers in the spinal cord and brain of the sea robin, *Prionotus carolinus* L.', *Journal of Comparative Neurology & Psychology*, 17, 307–327.

Atema, J. (1971), 'Structures and functions of the sense of taste in the catfish (*Ictalurus natalis*)', *Brain, Behavior and Evolution*, 4, 273–294.

Bardach, J.E. and Atema, J. (1971), 'The sense of taste in fishes', in Beidler, L.M. (ed.), *Handbook of Sensory Physiology*, Springer-Verlag, New York, pp. 293–336.

Caprio, J. (1975), 'High sensitivity of catfish taste receptors to amino acids', *Comparative Biochemistry and Physiology*, 52, 247–251.

Caprio, J. (1984), 'Olfaction and taste in fish', in Balis, L., Keynes, R.D., Maddrell, S.H.P. (eds), *Comparative Physiology of Sensory Organs*, Cambridge University Press, pp. 257–83.

Caprio, J. and Brand, J.G. (1993), 'The taste system of channel catfish: from biophysics to behavior', *TINS*, 16, 192–197.

Davenport, C.J. and Caprio, J. (1982), 'Taste and tactile recordings from the ramus recurrens facialis innervating flank taste buds in the catfish', *The Journal of Comparative Physiology [A]*, 147, 217–229.

Finger, T.E. (1983), 'The gustatory system in teleost fish', in Northcutt, R.G. and Davis, R.E. (eds), *Fish Neurobiology*, University of Michigan Press, Ann Arbor, MI, pp. 285–309.

Finger, T.E. and Morita, Y. (1985), 'Two gustatory systems: facial and vagal gustatory nuclei have different brainstem connections', *Science*, 227, (15 February), 776–778.

Kanwal, J.S. and Caprio, J. (1988), 'Overlapping taste and tactile maps of the oropharynx in the vagal lobe of the channel catfish, *Ictalurus punctatus*', *Journal of Neurobiology*, 19, 211–222.

Kanwal, J.S., Finger, T.E. and Caprio, J. (1988), 'Forebrain connections of the gustatory system in ictalurid catfishes', *Journal of Comparative Neurology*, 278, 353–376.

Kanwal, J.S., Hidaka, I. and Caprio, J. (1987), 'Taste responses to amino acids from facial nerve branches innervating oral and extra-oral taste buds in the channel catfish, *Ictalurus punctatus*', *Brain Research*, 406, 105–112.

http://news.nationalgeographic.com/news/2007/01/070116-manatees_2.htm

Finger, T. (1982), 'Somatotopy in the representation of the pectoral fin and free fin rays in the spinal cord of the sea robin, *Prionotus carolinus*', *Biological Bulletin*, 163, 154–161.

Insects and Spiders of the World (2003), Marshall Cavendish Corporation, ISBN: 978-076147-334-3.

Stirling, I. (1988), *Polar Bears*, University of Michigan Press, Ann Arbor, MI.

Stempniewicz, L. (2006), *Arctic*, 59, 247–251.

Chapter 5: Alarming Behaviour and Survival Strategies

AL: animal alarms and defence – pp. 276–314.

Novel electric signals in plants (using ion-selective micro-electrodes), htm, 9 March 2009, Max Planck Institute for Chemical Ecology.

Insect alarms, especially pheromones, are discussed throughout Holldobler, Bert and Wilson E.O., *The Superorganism* – the chemical communication quote is from p. 179.

Casey, C., 'Yellow Jackets in Season', *New York Times*, 23 July 2006; an article that specifically deals with the alarm pheromone and the researchers/entomologists who found/synthesised it: 'Science Watch: In a Hostile Environment, Drab Can Be Beautiful', *New York Times*, 10 November 1987.

Worm-grunting, fiddling, and charming: Catania, Kenneth C. (2008), 'Humans Unknowingly Mimic a Predator to Harvest Bait', Department of Biological Sciences, Vanderbilt University; PLoS *One* 3(10), e3472, doi:10:1371/journal.pone.0003472, 14 October 2008.

Vervet monkey warning calls research: Seyfarth, R.M., Cheney, D.L. and Marler, P. (1980), 'Monkey responses to three different alarm calls: evidence of predator classification and semantic communication', *Science*, 210 (4471), 801–803.

Owren, M.J. (1990), 'Acoustic classification of alarm calls by vervet monkeys (*Cercopithecus aethiops*) and humans (*Homo sapiens*)', *Journal of Comparative Psychology*, 104 (1), 20–28.

Templeton, C.N. and Greene, E. (2007), 'Nuthatches eavesdrop on variations in heterospecific chickadee mobbing alarm calls', *Proceedings of the National Academy of Sciences*, 104, 5479–5482.

'Chickadees Add Notes as Threat Grows', *Science News*, 25 June 2005.

Templeton, Christopher et al. (2005), 'Allometry of Alarm Calls: Black-Capped Chickadees Encode Information about Predator Size', *Science*, 308, 1934–1937.

Prairie dogs: Perla, B. and Slobodchikoff, C. (2002), 'Habitat structure and alarm call dialects in the Gunnison's prairie dog (*Cynomys gunnisoni*)', *Behavioral Ecology*, 13, 844–850.

Frederiksen, J.K and Slobodchikoff, C.N. (2007), 'Referential specificity in the alarm calls of the black-tailed prairie dog', *Ethology, Ecology and Evolution*, 19, 67–99.

Information on squirrel species and ultrasound: *Nature*, 1 July 2004.

Hot tails: Rundus, A., Owings, D., et al. (2007), 'Ground squirrels use an infrared signal to deter rattlesnake predation', *Proceedings of the National Academy of Sciences*, 19 July 2007.

Animals and regeneration: Nicolas Wade, 'Regrow Your Own', *Science Times, New York Times*, 11 April 2006.

Carlson, B.M. (2007), *Principles of Regenerative Biology*, Elsevier, London.

'Coyotes Have Arrived in the Eastern US', *Washington Post* magazine, 16 April 2006.

An excellent resource for understanding components of earthquakes is Joseph L. Kirschvink's report to the *Bulletin of the Seismological Society of America*, 90, 2:312–323, April 2000.

Homing pigeons are aware of ground tilting: Phillips, J.B. (1996), 'Magnetic Navigation', *Journal of Theoretical Biology*, 180, 309–319.

Chapter 6: From Frogcicles to Dreamstates

In a study by Layne and colleagues (Layne, J.R., Jr, Lee, R.E., Jr and Heil, T.L. (1989), *American Journal of Physiology*, 257, R1046–1049), within one minute of the onset of freezing, the heart rate of wood frogs (*Rana sylvatica*) nearly doubled to approximately eight beats per minute. The heart rate began to slow after the first hour of the freeze, and the heart completely stopped beating near the completion of ice formation approximately twenty hours later. Recordings from a single frog revealed that the heart beat resumes within one hour after thawing and near-normal function is achieved after only a few hours. The release of the latent heat of fusion caused a rise in body temperature (1.7°C) for a few hours and was closely correlated with an increase in the heart rate. However, other factors such as reduction in blood volume, increase in blood viscosity, and progressive hypoxia may prominently influence cardiac function indirectly. Regardless, the heart functions long enough to distribute glucose throughout the body during the first few hours of the freeze.

AL: hibernation – pp. 115–37.

Craig Heller's research on hibernation and sleep:

Von der Ohe, C., Heller, H.C. et al. (2007), 'Synaptic protein dynamics in Hibernation', *The Journal of Neuroscience*, 1 (27), 84–92.

Von der Ohe, C., Heller, H.C. et al. (2005), 'Ubiquitous and temperature-dependent neural plasticity in hibernation', *The Journal of Neuroscience*, 41 (26), 10590–8.

'Trying to Crack an Icy Mystery', an article by Matt Hood and Laura Stanton in the *Washington Post*, 12 December 2004, features the wood frog and spring peeper that go into a deep freeze during the cold weather.

'Lemur is First Known Hibernating Primate, Study Says', James Owen in England for *National Geographic News*, 23 June 2004.

An article about animals and regeneration: Nicolas Wade, 'Regrow Your Own', *Science Times*, *New York Times*, 11 April 2006.

Parrot fish and sleep: Videler, H. et al. (1999), 'Biochemical characteristics and antibiotic properties of the mucous envelope of the queen parrotfish', *Journal of Fish Biology*, 54, 1124–1127.

An article about animals and sleep: Zimmer, C., 'Down for the Count', *Science Times*, *New York Times*, 8 November 2005.

Good resources for animal questions, including horses and sleep: http.vetmed.illinois.edu; animals.nationalgeographic.com.

An update on research on sleep and animals, including genetic findings: Youngsteadt, E. (2008), 'Simple Sleepers', *Science*, 321, 334–337.

Rats dream about their tasks during slow-wave sleep: Lee, A.K. and Wilson, M.A. (2002), 'Memory of Sequential Experiences in the Hippocampus During Slow Wave Sleep', *Neuron*, 36, 1183–1194.

Chapter 7: Animal Marathons by Land and Sea

Two hypotheses have been proposed to account for turtles' remarkable navigation abilities. The first is that chemical cues emanating from target areas guide them to their destinations. This would be similar to the olfactory cues that homing salmon use to arrive at their spawning sites. The second is that turtles can approximate their position relative to target regions in the sea or on the shore using features of the earth's magnetic field.

AL: movement – pp. 89–99; migration – pp. 148–61.

An article on the research of turbulence theory and the flying (ballooning) spider appeared in the July 2006 issue of *BBSRC Business*. *Business* is the quarterly research highlights magazine of the UK's Biotechnology and Biological Sciences Research Council.

To read further about the monarch butterfly migrations:

'Internal Clock Leads Monarch Butterflies to Mexico', *National Geographic*, 16 February 2009.

Zhu, H., Sauman, I., Yuan, Q., Casselman, A., Emery-Le, M., Emery, P. and Reppert, S.M. (2008), 'Cryptochromes define a novel circadian clock mechanism in monarch butterflies that may underlie sun compass navigation', PLoS *Biology*, 6, e4.

Reppert, S.M. (2006), 'A colorful model of the circadian clock', *Cell*, 124, 233–236.

Sauman, I., Briscoe, A.D., Zhu, H., Shi, D., Froy, O., Stalleichen, J., Yuan, Q., Casselman, A. and Reppert, S.M. (2005), 'Connecting the navigational clock to sun compass input in monarch butterfly brain', *Neuron*, 46, 457–467.

Gugliotta, G., 'Butterflies Guided by Body Clocks: Sun Scientists Shine Light on Monarchs' Pilgrimage', *Washington Post*, 23 May 2003, p. A03.

An article by Cornelia Dean in the *New York Times National*, 13 February 2009, 'Charting Bird Migrations by Using Tiny Backpacks', explains Bridget Stutchbury's newly published research in *Science*, 323, 896–898.

The unique flying skills of hummingbirds are featured in a *Washington Post Science* article: 'Supreme Fliers Share a Shortcut for Stability', by Joel Achenbach, 13 April 2009.

Chapter 8: Wit, Wiles and Good Fun

AL: animal intelligence – pp. 471–85.

Chris Counts describes the research of field biologist bat researcher Matina Kalcounis-Rüppell which picked up the singing of local deer mice with ultrasonic recording equipment in '"Mouse Idol" in Carmel Valley', in *The Carmel Pine Cone*, 21 April 2006.

Rats have been shown to chirp delightedly above the range of human hearing when wrestling with each other or being tickled by a keeper – the same vocalisations they make before receiving morphine or having sex.

For a demonstration of rat laughter, see: http://www.youtube.com/watch?v=C0kxmfSGCaE

Studying sounds of joy may help us understand the evolution of human emotions and the brain chemistry underlying such emotional problems as autism and attention-deficit/hyperactivity disorders, says Jaak Panksepp, a pioneering neuroscientist who first discovered rat laughter when animals are tickled by the human hand. The researchers also found that rats would rather spend time with animals that chirp a lot than with those that don't. Panksepp said that laughter, at least in response to a direct physical stimulus such as tickling, may be a common trait shared by all mammals.

Panksepp, J. and Burgdorf, J. (1999), 'Laughing rats? Playful tickling arouses high frequency ultrasonic chirping in young rodents', in Hameroff, S., Chalmers, D. and Kazniak, A. (eds), *Toward a Science of Consciousness*, Vol. 3, MIT Press, pp. 124–36.

Panksepp, J. and Burgdorf, J. (2003), '"Laughing" rats and the evolutionary antecedents of human joy?', *Physiology and Behavior*, 79, 533–547.

The title of this book is based on anecdotal evidence that mice too respond warmly to the human touch and make high-frequency 'giggling' sounds when tickled.

Two young scientists, Roian Egnor at the Howard Hughes Medical Institute, Janelia Farm, and Christine Portfors at Washington State University, Vancouver, are using modern technology to study social communication in mice. They observe that male and female mice vocalise profusely with their cage-mates during a reunion after a long-term separation (personal communication), as is the case with moustached bats. We take these to be sounds of 'joy' or affection (and/or for re-establishing dominance) as expressed by human infants when reunited, after a relatively long-term separation, with a parent or sibling.

Brown, Stuart, *Play*, Avery, 2009 (bear/dog play quote: pp. 21–4).

Marc Beckoff and John Byers' book on animal play (*Animal Play: evolutionary, comparative, and ecological perspectives* (1998), Cambridge University Press) goes into more details about play in a wide range of species.

Information about Alex the African grey parrot is from the book *Alex and Me* by Irene Pepperberg (2008); Toby the African grey parrot: March 2009 personal conversation with Richard Restak.

Relative to body size, big-brained birds such as parrots and crows live longer than smaller-brained species such as grouse and pigeons: Sol, D. et al. (January 2007), *Proceedings of the Royal Society B*.

Primate research anecdotes: March 2009 personal conversation with Karl Pribram.

Mulcahy, N. and Call, J. (2006), 'Apes Save Tools for Future Use', *Science*, 312, 1038–1040.

There's a wonderful research summary and photographs of the memory research done in Japan on BBC News (3 December 2007), 'Chimps Beat Humans in Memory Test', by Helen Briggs.

Susan Milius writes in *Science News*: 'Chimp Champ: Ape Aces Memory Test, Outscores People', 172, 23 (8 December 2007), 355; 'Furry Math: Macaques Can Do Sums Like People in a Hurry', 172, 28 (22 December 2007), 390; 'Honeybee Tells 2 from 3', 27 January 2009.

Claudia Uller of the University of Essex presents a thorough study of numerical cognition in the *Journal of Evolutionary Psychology*, 6 (2008), 4, 237–253 (pdf).

New research has given credence to the idea that laughter evolved in a common ancestor of the great apes and humans. For a demonstration, see: http://news.bbc.co.uk/2/hi/science/nature/8083230.stm

Chapter 9: Eavesdropping and Deception

AL: deception and animals – pp. 210, 250, 296–311, 472.

Ecologists report in *Science* how 'Butterfly Larvae Trick Ants with Scent and Sound': www.npr.org

Brock Fenton, M. (1985), *Communication in Chiroptera*, Indiana University Press.

Page, R.A. and Ryan, M.J. (2005), 'Flexibility in assessment of prey cues: frog-eating bats and frog calls', *Proceedings of the Royal Society B*, 22, Vol. 272, no. 1565, 841–847; calls: doi: 10.1098/rspb.2004.2998.

McGregor, Peter K. (2005), *Animal Communication Networks*, Cambridge University Press.

Wilson, Edward O. (2000), *Sociobiology: The New Synthesis*, Harvard University Press.

Balcombe, J.P. and Fenton, M.B. (1988), 'Eavesdropping by Bats – the Influence of Echolocation Call Design and Foraging Strategy', *Ethology*, 79, 158–166.

Fenton, M.B. (2003), 'Eavesdropping on the echolocation and social calls of bats', *Mammal Review*, 33, 193–204.

Marler, P., Dufty, A. and Pickert, R. (1986a), 'Vocal communication in the domestic chicken: I. Does a sender communicate information about the quality of a food referent to a receiver?', *Animal Behavior*, 34, 188–193.

Marler, P., Dugty, A. and Pickert, R. (1986b), 'Vocal communication in the domestic chicken: II. Is a sender sensitive to the presence and nature of a receiver?', *Animal Behavior*, 34, 194–198.

Mitchell, R.W. and Anderson, J.R. (1997), 'Pointing, withholding information, and deception in capuchin monkeys (*Cebus apella*)', *Journal of Comparative Psychology*, 111, 351–361.

Pearl, D.L. and Fenton, M.B. (1996), 'Can echolocation calls provide information about group identity in the little brown bat (*Myotis lucifugus*)?', *Canadian Journal of Zoology – Revue Canadienne de Zoologie*, 74, 2184–2192.

Chapter 10: Rhythm, Song and Dance

AL: sounds, song and music – pp. 456–61.

Bispham, J. (22 August 2006), 'Rhythm In Music: What Is It? Who Has It? And Why?', University of Cambridge, Bispham-pdf4(1).pdf.

Singing is used to establish roosting territories in the sac-winged bat, which employs an unusually large vocal repertoire. In this species, males emit tonal calls while interacting with females and other types of calls consisting of composites when actively defending their territories. Their songs consist of short repeated tones that do not appear to have any obvious context other than advertising the quality of singing by a male.

http://www.pbs.org/lifeofbirds/songs/

Ivan Turk of the Slovenian Academy of Sciences in Ljubljana discovered a hollowed-out bear bone pierced on one side with four complete or partial holes as the earliest known musical instrument. The perforated bone, found in an Eastern European cave, represents a flute made and played by Neanderthals at least 43,000 years ago, the scientists contended. Archaeologists in China found what is believed to be the oldest still-playable musical instrument: a 9,000-year-old flute carved from the wing bone of a crane. For additional details on this topic go to the musicologist Bob Fink's website: http://www.greenwych.ca/2001-2.htm

Experimental evidence for synchronisation to a musical beat in a non-human animal:

Patel, A.D., Iversen, J.R., Bregman, M.R. and Schulz, I. (2009), *Current Biology*, Elsevier.

Information theory is based on a mathematical study of data encoding and transmission. It provides a quantitative analysis of the complexity and structure of the whale songs. Information theory allows one to study the structure of humpback songs without knowing what they mean.

Singing in bats was first reported by Vaughan (1976) in *Cardioderma cor*, an African *megaderma* bat species.

To gain a deeper understanding about the rituals of the dawn chorus, we have to dig deeper into the dynamics of group structure and the inner workings of the brain. One of the hormones that is known to play a critical role in social bonding is oxytocin. Oxytocin is released in our brains with every romantic kiss and even when a group of women are simply talking to each other in a social setting. Interestingly, oxytocin was also found to be present in the neurons throughout the auditory system of the brain in bats, and this may be the case in other species, including humans. This means that simply hearing others talking in a social context may stimulate auditory neurons to manufacture and release oxytocin. What's more, oxytocin is also released when our stomachs are full. Now imagine a group of people or animals having a good breakfast or dinner and chattering away. What they are really doing is flooding their brains with oxytocin and creating a social bond that can last at least until the next day – and in the case of prairie voles, for their entire lifetime.

Oxytocin is made of a chain of just nine amino acids (a nonapeptide) and is a hormone that is released into the blood by the pituitary during parturition and the let-down of milk. Oxytocin is also a neurohormone that is secreted by neurons in the hypothalamus rather than the endocrine cells in a gland, such as the pancreas or the thyroid. More recently it has been shown that oxytocin is also a neurotransmitter that is widely present in the brain, and the presence or absence of oxytocin receptors determines whether a species of prairie voles is going to be monogamous or polygamous. It has been labelled as the hormone for social bonding.

Chapter 11: Flirting, Courting and Coupling

AL: pheromones – pp. 444–8; sex and reproduction – pp. 322–75.

Bass, A.H., Bodnar, D.A. and Marchaterre, M.A. (2000), 'Midbrain acoustic circuitry in a vocalizing fish', *Journal of Comparative Neurology*, 419, 505–531.

Christopher Clark of the University of California, Berkeley and colleagues found that the male Anna's hummingbird emits a courting chirp with its tail which is almost identical to his vocal song. The tail feathers functioned like musical reeds, something previously unknown in birds. The study was published online, 29 January 2008, in the *Proceedings of the Royal Society B*.

About the poor dibbler: Dickman, C. (January 2008), *Fragile Balance: The Extraordinary Story of Australian Marsupials*, University of Chicago Press.

Komodo dragon virgin birth at Chester Zoo, 21 December 2006: science.msnbc.com, id:16298548. As for the long-tailed manakin, the link to the actual video is: www.uwyo.edu/dbmcd/lab/LTMvideo.htm

Brown University researchers discover stuttering in frogs: Suggs, D.N. and Simmons, A.M. (2005), 'Information theory analysis of pattern of modulation in the advertisement call of the male bullfrog, *Rana catesbeiana*', *Journal of the Acoustical Society of America*, 117 (4 Pt. 1), 2330–2337.

Cryptochromes trigger mass spawning of corals: Levy, O., et al. (2007), 'Light-Responsive Cryptochromes from a Simple Multicellular Animal, the Coral *Acropora millepora*', *Science Magazine*, October 2007, Vol. 318 (5849), 467–470.

Different aspects of the black and white markings of the male lark bunting are preferred by females in different years: Chaine, A. and Lyon, B. (25 January 2008), 'Adapative Plasticity in Female Mate Choice Dampens Sexual Selection on Male Ornaments in the Lark Bunting', *Science*, Vol. 319, 459–462.

Jacobson's organ, pheromones and giraffe reproductive behaviour: Margulis, J. (2008), 'Looking Up', *Smithsonian Magazine*, November 2008, 36–42.

Francis Champagne of Columbia University explores the genetic and environmental factors that may determine how maternal behaviours are passed from mothers to daughters (oestrogen-oxytocin): Champagne, F.A., Weaver, I.C., Diorio, J., Dymov, S., Szyf, M. and Meaney, M.J. (in press), 'Maternal care associated with methylation of the estrogen receptor alpha 1b promoter and estrogen receptor alpha expression in the medial preoptic area of female offspring', *Endocrinology*.

Levin, A. (2009), 'Early Experiences Change DNA and Thus Gene Expression', *Psychiatric News*, 5 June 2009, Vol. 44, No. 11, p. 18.

Donaldson, Z.R. and Young, L.J. (2008), 'Oxytocin, Vasopressin, and the Neurogenetics of Sociality', *Science*, Vol. 322, 7 November 2008, 900–904.

Goleman, Daniel (2006), *Social Intelligence*, Bantam Dell, pp. 202–03.

Index